Over, Around, and Within

Geometry and Measurement

STUDENT BOOK

TERC

Mary Jane Schmitt, Myriam Steinback,
Tricia Donovan, and Martha Merson

Bothell, WA • Chicago, IL • Columbus, OH • New York, NY

TERC
2067 Massachusetts Avenue
Cambridge, Massachusetts 02140

EMPower Research and Development Team
Principal Investigator: Myriam Steinback
Co-Principal Investigator: Mary Jane Schmitt
Research Associate: Martha Merson
Curriculum Developer: Tricia Donovan

Contributing Authors
Donna Curry
Marlene Kliman

Technical Team
Graphic Designer and Project Assistant: Juania Ashley
Production and Design Coordinator: Valerie Martin
Copyeditor: Jill Pellarin

Evaluation Team
Brett Consulting Group:
   Belle Brett
   Marilyn Matzko

EMPower™ was developed at TERC in Cambridge, Massachusetts. This material is based upon work supported by the National Science Foundation under award number ESI-9911410. Any opinions, findings, and conclusions or recommendations expressed in this publication are those of the authors and do not necessarily reflect the views of the National Science Foundation.

TERC is a not-for-profit education research and development organization dedicated to improving mathematics, science, and technology teaching and learning.

All other registered trademarks and trademarks in this book are the property of their respective holders.

http://empower.terc.edu

Printed in the United States of America
6 7 8 9 10   QTN   20 19 18 17 16

ISBN 978-0-07662-089-0
MHID 0-07-662089-1

Contents

Introduction

Welcome to EMPower

Students using the EMPower books often find that EMPower's approach to mathematics is different from the approach found in other math books. For some students, it is new to talk about mathematics and to work on math in pairs or groups. The math in the EMPower books will help you connect the math you use in everyday life to the math you learn in your courses.

We asked some students what they thought about EMPower's approach. We thought we would share some of their thoughts with you to help you know what to expect.

"It's more hands-on."

"More interesting."

"I use it in my life."

"We learn to work as a team."

"Our answers come from each other… [then] we work it out ourselves."

"Real-life examples like shopping and money are good."

"The lessons are interesting."

"I can help my children with their homework."

"It makes my brain work."

"Math is fun."

EMPower's goal is to make you think and to give you puzzles you will want to solve. Work hard. Work smart. Think deeply. Ask why.

Using This Book

This book is organized by lessons. Each lesson has the same format.

- The first page explains the lesson and states the purpose of the activity. Look for a question to keep in mind as you work.

- The activity page comes next. You will work on the activities in class, sometimes with a partner or in a group.

- Look for shaded boxes with additional information and ideas to help you get started if you become stuck.

- Practice pages follow the activities. These practices will make sense to you after you have done the activity. The three types of practice pages are

  🎯 *Practice*: Provides another chance to see the math from the activity and to use new skills.

  ⊘ *Extension*: Presents a challenge with a more difficult problem or a new but related math idea.

  Test Practice: Asks a number of multiple-choice questions and one open-ended question.

In the *Appendices* at the end of the book, there is space for you to keep track of what you have learned and to record your thoughts about how you can use the information.

- Use notes, definitions, and drawings to help you remember new words in *Vocabulary*, pages 149–151.

- Answer the *Reflections* questions after each lesson, pages 152–159.

Check *Sources and Resources*, page 160, for books and Internet sites related to the unit.

Tips for Success

Where do I begin?

Many people do not know where to begin when they look at their math assignments. If this happens to you, first try to organize your information.

Much of this unit is about angles, lines, shapes, and their measurements.

Use your powers of observation.

Ask yourself:

> *What shapes or parts of shapes do I see?*
>
> *How do the parts relate to one another? Can a shape be broken into other shapes or parts I already know?*

A drawing or a real object always helps. If you are unsure of a word, look it up in a dictionary or ask a friend, then make a sketch to go along with it.

Ask yourself:

> *Can I draw what the words are saying? Or, can I find an object around the house that is described in the problem?*

Another part of getting organized is figuring out what skills are required.

Ask yourself:

> *What aspect of the shape should I focus on—the angles, the sides, the distance around, the surface, or the space within a shape?*
>
> *What measurements do I know?*

Write down what you already know.

I cannot do it. It seems too hard.

Make the numbers smaller. Deal with just a little bit of data at a time. Cross out information or data that you do not need.

Ask yourself:

Is there information in this problem that I do not need?

Have I ever seen something like this before? What did I do then?

You can always look back at another lesson for ideas.

Am I done?

Don't walk away yet. Check your answers to make sure they make sense.

Ask yourself:

Do the words and drawings match well? Does the result seem logical?

Did I answer the question? (Always read it one more time!)

Check your math with a calculator. Ask others whether your work makes sense to them.

Opening the Unit:
Geometry
Groundwork

*Is there geometry
in your life?*

Often when we are making something, ideas associated with the study of geometry arise. Do you see shapes within shapes? How do parts fit together? How does measuring help you figure things out when you are cooking, building, or sewing?

In this unit, you will learn about shape, line, and form. You will explore the skills that allow people to work mathematically with shapes in everyday life.

Activity 1: Making a Mind Map

Make a Mind Map using words, numbers, pictures, or ideas that come to mind when you think of *geometry* and *measurement*.

GEOMETRY

MEASUREMENT

Activity 2: Making an Angle Demonstrator

Make your own angle demonstrator with the parts that your teacher will give you.

- Cut out the two strips of paper and place one on top of the other so the two dots are aligned.

- Put a fastener through the two dots so the paper strips can open and close like a mouth.

Now you have an angle demonstrator. Make another one for home use, if you like.

Activity 3: Initial Assessment

Your teacher will show you some problems and ask you to check off how you feel about your ability to solve each problem:

___ Can do   ____ Don't know how   ____ Not sure

Have you ever noticed that every new place you work has its own words or specialized vocabulary? This is true of topics in math too. In every lesson you will be introduced to some specialized vocabulary. Do not worry if you see words in the problems that you do not recognize. You can write some words down and look them up later, or learn as you go.

Activity 4: Homemade Objects

Choose an object from the collection provided, or use a homemade object of your own.

Object: _____

1. Sketch your object.

Activity 4: Homemade Objects *(continued)*

2. What shapes make up your object? Make a sketch of each shape you see. Name each shape.

Sketch Sheet

Sketch of Shapes Observed	Name or Description of Shapes

3. When you finish drawing and naming the shapes, trade your object and Sketch Sheet with a classmate. See if you can find any additional shapes in your partner's object. Add them to your partner's Sketch Sheet and name them.

Practice: Using Geometry

Write three paragraphs about how you use geometry and measurement.

- In your work...

- At home...

- In everyday life in your community...

Practice: Seeing Geometry

1. Sketch an object that has some interesting shapes.

2. What shapes and angles do you see in the object?

3. What measurements could you take of the object?

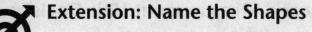

Extension: Name the Shapes

The following are traditional Japanese patterns.

1. Sketch and name the shapes you see within this pattern.

2. Sketch and name the shapes you see within this pattern.

3. Many countries have traditional designs that include geometric shapes.

   a. Research a design from your own or another tradition.

   b. Sketch the design.

   c. Name the shapes you see within the design.

Geometry and Measurement Unit Goals

- Recognize and describe shapes and their characteristics.

- Find area and perimeter of rectangles and volume of rectangular solids.

- Make drawings to scale.

- Use linear, square, and cubic units.

- Use spatial reasoning to solve problems.

- Make generalizations about two- and three-dimensional shapes.

My Own Goals

Sharing Secret Designs

What shapes surround you?

When you watch a movie or read a book, you get to know the people in it, the characters. These people have their own set of features or characteristics that set them apart from everyone else. The same is true of shapes. In this lesson, you will learn the characteristics that make shapes unique, or different from each other.

Geometry has its own language. You will begin to understand geometry terms and keep track of them by adding new words and their definitions to your vocabulary list.

Activity 1: Guess My Shape

Student holding the shape, tell your classmate

- The number of angles;

- The number of sides;

- The relationship of the sides to each other. (Do they touch? How?)

Student guessing the shape

- What do you think the shape looks like?

- Sketch it.

- Do you know a name for this shape?

- Would your name or sketch change if you held the shape upside down? What if you held it sideways?

Shape Description Chart

Basic Geometric Shapes	Where I See This Shape	Shape Characteristics	Shape Names
1. ▢			
2. ◺			
3. ▯			
4. ▱			
5. ◿			
6. △			

Basic Geometric Shapes	Where I See This Shape	Shape Characteristics	Shape Names
7.			
8.			
9.			
10.			
11.			
12.			

Activity 2: Sharing Secret Designs

Goals

- Use geometry terms to guide your partner to make the exact design you made.

- As you work, think about the shapes and how they fit together.

1. **Making a design**

   - Use no more than six shapes.

   - Use at least three different shapes.

   - Do not let your partner see your design.

2. **Giving directions**

   - Give directions to your partner one statement at a time.

   - Do not use color names to identify shapes; use only the names of the shapes.

3. **Getting directions**

   - Ask only yes/no questions to clarify directions. Example: "Does the short side of the **trapezoid** connect to the hexagon?"

4. **Checking your work**

   - When you think you know your partner's design, ask, "Is this your design?"

   - If your partner answers "yes," switch roles; it is your partner's turn to guess your design.

   - If your partner answers "no," get more clues.

5. **Starting over again with a new designer**

   - When you and your partner have both guessed each other's designs, find new partners.

   - Repeat the steps above.

6. **Making a list**

   - Write down the words you used most often to describe the shapes and the designs.

> Be sure to give good directions. Use specific geometry terms.

Practice: Road Signs Match

Write the name of each shape on the top line next to it. Then write the letter of the road sign on the bottom line next to the shape it best matches.

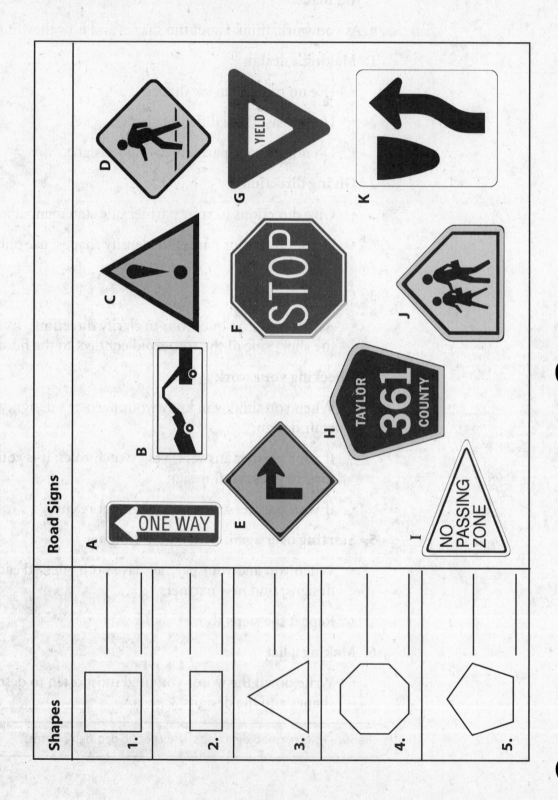

Practice: Alike Yet Different

Look at the shapes below, and answer the following questions. Support your ideas with sketches and explanations.

1. How are these two shapes alike?

2. How are they different?

3. How would you describe the angles in each shape?

Practice: Covering Hexagons

Cover the **hexagons** (six-sided shapes) on the following page with any combination of these three smaller shapes from the Pattern Blocks.

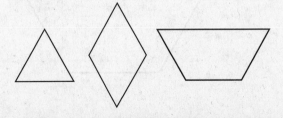

If you have Internet access, go to http://arcytech.org/java/patterns. This Web site allows you to move the pieces around on your screen.

Sketch the ways you used the shapes to cover the hexagons.

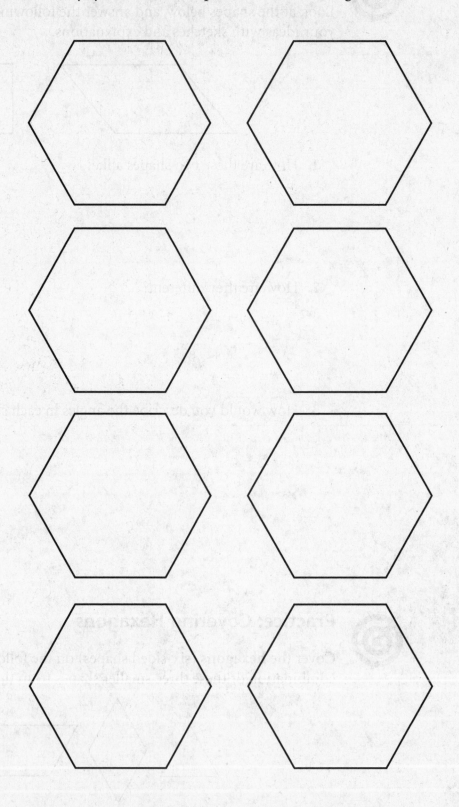

Extension: Hunt for Shapes

Choose Problem 1 or 2 to do.

1. Take a walk outside, in a room where you live, or in a store, and find as many shapes as you can.

   a. Where did you go?

   b. What shapes did you see? Name them, and draw them or take pictures of them with a camera.

2. Look through a magazine.

   a. Cut out different shapes, and glue them on a piece of paper or put them in an envelope.

   b. What shapes did you find? Name them.

Test Practice

Items 1–6 are based on the design below.

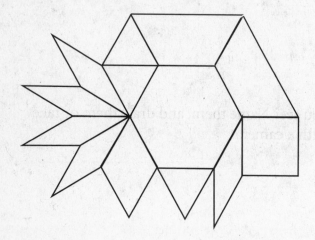

1. The name of the largest shape in this design:

   (1) Octagon

   (2) Square

   (3) Triangle

   (4) Hexagon

   (5) Rectangle

2. The most frequent shape in this design:

   (1) Trapezoid

   (2) Diamond, or rhombus

   (3) Triangle

   (4) Square

   (5) Octagon

3. How many shapes share a side with the hexagon?

   (1) 2

   (2) 3

   (3) 5

   (4) 6

   (5) 8

4. The number of triangles in the design:

   (1) 1

   (2) 2

   (3) 3

   (4) 4

   (5) 5

5. The square can be recognized by

   (1) Three sides.

   (2) Pointy corners.

   (3) Five equal sides.

   (4) Four parallel sides.

   (5) Two sets of equal, parallel sides.

6. Of the 13 shapes in the design, how many are four-sided shapes?

Get It Right

At what angle is reading comfortable?

In this lesson, you will look for angles and then work with **right angles**, also called 90-degree angles. Right angles are formed by two **perpendicular lines.**

You will become very familiar with right angles. By using the right angle as a benchmark, you can estimate the size of many other angles. It is essential to measure angles correctly when you are building anything that needs to be level or stand up straight. If the measure is off, the sides will not align correctly. What you are creating could be as small as a cookie cutter or as large as a bridge. Imagine the results if the sides do not align!

Activity: What Is a Comfortable Angle?

All people like to hold a book at the same angle to read it—or do they?

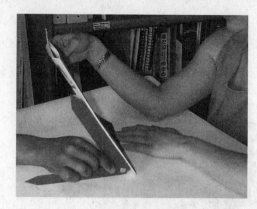

1. Place a clipboard flat on the table in front of you.

   Sit straight in your chair. Do not lean forward or drop your head.

2. Tilt the clipboard up to an angle where it is easy for you to read.

3. Your partner measures and traces the angle, using an angle demonstrator. Your partner should

   - Put one side of the tool on the desk;

   - Place the other side against the clipboard;

   - Trace the angle on a blank sheet of paper;

   - Mark the traced angle "∠A."

4. Keep tilting the clipboard toward you until it is impossible to read.

   Your partner measures and traces the angle and marks it "∠B."

5. Change places and record the angles for your partner.

6. Answer the questions on the next page.

Questions

1. Describe what happened to the angle as you tipped the clipboard.

2. Compare your angles with those of other students. What do you notice? Are their angles smaller than yours, larger, or about the same?

3. Use a protractor to measure ∠A and ∠B. Write the degree measurement next to each angle.

4. Compare the degree measurements of your ∠A and ∠B with those of two classmates. Describe how they are different.

5. What conclusions can you draw about reading angles?

Practice: Sign on the Line

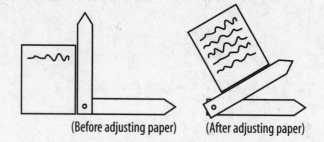

(Before adjusting paper)   (After adjusting paper)

Place a piece of paper flat on your desk at an angle at which it is comfortable for you to write. Line up one side of your angle demonstrator with the bottom edge of the desk, and move the other side to line up with the bottom edge of the paper.

1. Trace the angle.

2. Estimate the angle's size. _____

3. Measure the angle with a protractor. _____

4. How does your angle measurement compare with the angle in the picture above?

Practice: Less Than, Greater Than, or Equal to 90°

1. Label each of the following angles **obtuse** (greater than 90°), **acute** (less than 90°), or **right** (equal to 90°). Mark each angle with an arc or the right angle mark. Estimate the size. Write the estimate in degrees.

A._____   B._____   C._____

D._____   E._____   F._____

2. Refer to the figure below.

a. How many angles do you see in the large K?

b. Estimate the size of each angle, and record your estimates.

3. Look at the following figure of a person at a podium.

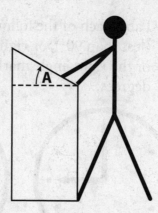

  a. Label the angles you see.

  b. List the angles and estimate their sizes.

  c. Find the exact measurement for Angle A.

4. Sketch three angles: a right angle, an obtuse angle, and an acute angle. Label each sketch.

Practice: Shapes and Angles

Complete the table.

Shape	Type of Angle	Angle's Measure (Use a protractor.)
1.	∠A _____ ∠B _____ ∠C _____	
2.	∠A _____ ∠B _____ ∠C _____ ∠D _____	
3.	∠A _____ ∠B _____ ∠C _____ ∠D _____ ∠E _____ ∠F _____	
4.	∠A _____ ∠B _____ ∠C _____ ∠D _____	

5. Was there anything that you noticed about the angle measurements of the various shapes? If so, what did you discover?

Extension: 90°

You know that a right angle measures 90°. Use that knowledge and your arithmetic skills to figure out the measurements in degrees of the missing angles in the figures below.

When you are done, sketch two of your own examples.

Show your work for Problems 2–5. How did you determine the measure(s) of the missing angle(s)?

1.

∠A = _____

2.

∠A = 45°   ∠B = _____

3.

∠A = _____   ∠B = 60°

4.

∠A = 10°   ∠B = ∠C

∠B = _____   ∠C = _____

5.

∠A = _____   ∠B is twice as large as ∠C

∠B = _____   ∠C = _____

Draw two examples of your own in the space below.

6.

7.

Extension: Degrees of Comfort

Choose one of the following problems to do:

1. Describe a real-life situation where a change in angle would affect how well something worked.

or

2. Draw a design for a reading stand. Use a protractor to measure and label all your angles.

Test Practice

1. Which of the following is the correct measurement for a right angle?

   (1) 45°

   (2) 60°

   (3) 90°

   (4) 180°

   (5) 360°

2. Which expression below shows how to find the measure of Angle B?

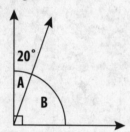

   (1) 90(20)

   (2) 90 − 20

   (3) 20 − 90

   (4) 90 + 20

   (5) 90 ÷ 20

3. Which of the following is *not* an example of a right angle?

   (1) The corner of a piece of paper

   (2) The corner of a book

   (3) The corner of a stop sign

   (4) The corner of a door

   (5) The corner of a tissue box

4. Which of the following angles measure(s) less than 90°?

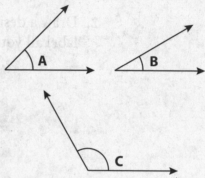

   (1) Angle A

   (2) Angle B

   (3) Angle C

   (4) Angles A and B

   (5) Angles B and C

5. The hands of the clock are at a right angle when the time is

   (1) 1:00 p.m.

   (2) 3:00 a.m.

   (3) 5:00 a.m.

   (4) 7:00 p.m.

   (5) 8:00 a.m.

6. What is the total number of degrees in the rectangle below?

Get It Straight

> *Where do you see straight angles?*

In this lesson, you will study **straight angles**. Most of us reach for a ruler to make a straight line. Construction workers use a tool called a level to make sure the walls they build are straight, but how does anyone know for certain what "straight" is?

You will identify straight angles in this lesson and use that information to find the measure of other types of angles.

Activity 1: Angles in Triangles

Your teacher will give you a triangle.

1. Name the triangle by its type: isosceles, right, equilateral, or scalene.

2. Label the three angles of the triangle, tear them out, and tape them onto the line below.

3. What is the total measurement of the angles of the triangle?

Activity 2: On the Line

Station 1: Triangles

1. Choose one of the triangles.
   Record the three angle labels: _____

2. Tear off the angles. Tape them on the line on page 30 with all the labeled angles touching one another.

3. What is the sum of the angles of your triangle? _____

Station 2: Rectangles

1. Choose one of the rectangles.
   Record the four angle labels: _____

2. Tear off the angles. Tape them on the line on page 30 with all the labeled angles touching one another.

3. What is the sum of the angles of your rectangle? _____

Station 3: Personal Triangle

1. Make your own triangle—any size you like.

2. Trace it so you have another copy.

3. Cut out the triangle on one copy, and label each angle.

4. Tear off the angles. Tape them on the line on page 30.

5. What is the sum of the angles? _____

Station 4: At the X

1. Measure one of the four angles of the X with your protractor.

2. Figure out the measures of the other three angles based on your one angle measurement and what you know about straight lines.

3. Draw the X below, label the angles, and record their measurements in degrees.

4. Check your measurements using a protractor.

Station 5: Triangle Types

In geometry there are four types of triangles:

- **Right triangles** have one right angle.

- **Isosceles triangles** have two equal angles and two equal sides.

- **Equilateral triangles** have three equal angles and three equal sides.

- **Scalene triangles** have no equal angles or sides.

1. Look at the cards. Sort the triangles by type: right, isosceles, equilateral, or scalene. Include the triangle you made in Station 3.

2. Write the triangle numbers underneath each heading name below.

3. If a triangle fits more than one name, you can write its number in more than one space.

4. Cover your answers by taping a paper strip over them, and shuffle the cards before you leave.

Type of Triangle	Right	Isosceles	Equilateral	Scalene
Card Number				

Practice: Missing Angle Measures

1. Make a guide for yourself.

   a. Right ∠ = _____

   b. Straight ∠ = _____

   c. All ∠s in a triangle total _____

2. Find the measure(s) of the missing angle(s) in each triangle below. Show how you figured out the measure(s).

a.

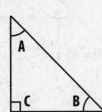

∠A = 45°

∠B = _____

∠C = _____

b.

∠A = 20°
∠B = 10°

∠C = _____

c.

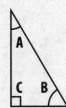

∠A = 30°

∠B = _____

∠C = _____

d.

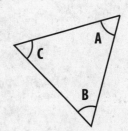

∠A = _____

∠B = 60°

∠C = 60°

3. Find the measures of the missing angles in the square below. Show how you figured out the measures.

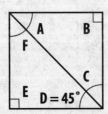

∠A _____   ∠D ___45°___

∠B _____   ∠E _____

∠C _____   ∠F _____

Practice: Angles and Roads

Imagine your job is to do road layouts for the town.

You do not want to measure every angle at every intersection, so you measure one and then figure out the others.

What are the angles in these four road layouts? Show your work.

1.

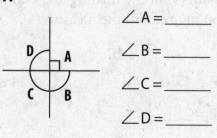

∠A = _____

∠B = _____

∠C = _____

∠D = _____

2.

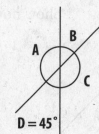

D = 45°

∠A = _____

∠B = _____

∠C = _____

∠D = _____

3.

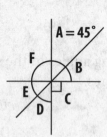

A = 45°

∠A = _____

∠B = _____

∠C = _____

∠D = _____

∠E = _____

∠F = _____

4.

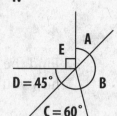

D = 45°

C = 60°

∠A = _____

∠B = _____

∠C = _____

∠D = _____

∠E = _____

Challenge Problems

Find the measures of each of the angles in Problems 5 and 6.

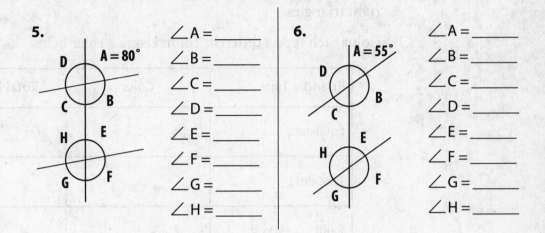

5.

∠A = _____
∠B = _____
∠C = _____
∠D = _____
∠E = _____
∠F = _____
∠G = _____
∠H = _____

6.

∠A = _____
∠B = _____
∠C = _____
∠D = _____
∠E = _____
∠F = _____
∠G = _____
∠H = _____

Did you notice any patterns? If you did, why do you think that is so?

7. Draw another road intersection. Label your angles with letters. Use a protractor to establish one of your angle measures. Then use what you know about angles to figure out the measures of the others.

Practice: Quilted Triangles

In the crazy quilt below, you will find equilateral, isosceles, scalene, and right triangles.

Color each type a different color. Using a ruler helps!

Triangle Type	Color	Total Number
Equilateral		
Isosceles		
Right		
Scalene		
Isosceles Right		

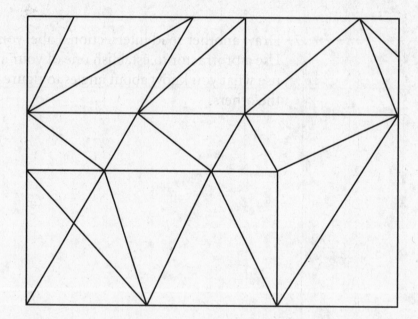

Practice: Measure It Up

1. Use a protractor where needed to measure the angles in the triangles below.

2. Write the measurement next to each angle.

3. Identify as many triangles as you can: equilateral, scalene, isosceles, right.

A.

B.

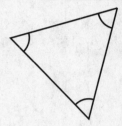

C.

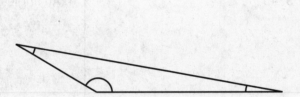

D.

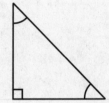

E.

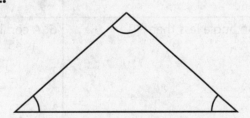

4. Check your work with the protractor by figuring the sum of the angle measures for each triangle. What is the sum of each?

Extension: Hunting for Angles in the Real World

- Find an example in real life of each of the angle measures. You will need to find six examples in all.

- Identify the object in which you found the angle.

- Sketch the angle.

- If you can, sketch the object. Use arrows and angle labels to show where you saw the angle in the object.

1. 90°	2. 180°
3. 45°	**4.** An angle greater than 90°
5. An angle less than 90°	**6.** A combination of angles, such as 45° and 90° or 45° and 180°

Test Practice

1. Which of the following is the correct measurement for a straight angle?

   (1) 45°

   (2) 90°

   (3) 100°

   (4) 180°

   (5) 360°

2. Which of the following angle combinations would result in a 180° angle?

   (1) A right angle and a 60° angle

   (2) Three right angles

   (3) A right angle and a 100° angle

   (4) A right angle, a 60° angle, and a 20° angle

   (5) A right angle and two 45° angles

3. Which of the following expressions represents the measure of ∠M?

   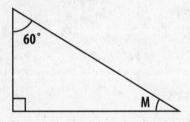

   (1) 180 − (90 + 60)

   (2) (90 + 60) − 180

   (3) 180 − 90 + 60

   (4) 90 − 60 + 180

   (5) 180 − (90 − 60)

4. In an equilateral triangle

   (1) All angles are 60°.

   (2) There is one right angle.

   (3) Two angles are 45°.

   (4) The sum of the three angles is 185°.

   (5) One angle is greater than 60°.

5. In the diagram below, you know that Angle A is not a right angle, Angles A and B total 180°, and Angles B and C total 180°. What else is also true?

   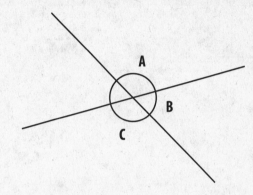

   (1) ∠A is equal to ∠B.

   (2) ∠A is equal to ∠C.

   (3) ∠A is twice as large as ∠B.

   (4) ∠C is equal to ∠B.

   (5) ∠C is twice as large as ∠B.

6. What is the measure of ∠P in the triangle?

   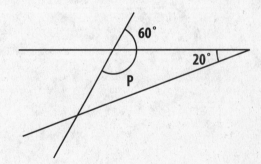

Giant-Size

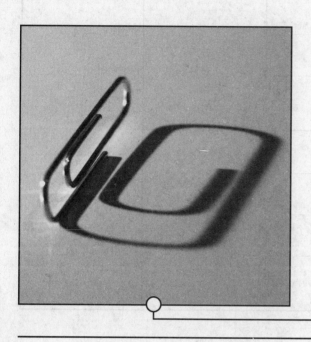

> *How do the shapes compare?*

In this lesson, you will explore **similar shapes**. In what ways can a shape change yet still be similar to the original shape? For instance, if you enlarge a triangle six times, how is the new shape similar to the first one? What if you lengthen only one side of the same triangle? Is the new triangle similar to the original? How will these changes affect the **perimeter**?

You will learn a rule for keeping shapes similar no matter how large or small you make them.

Activity: Giant-Size

Name of Item	Original Shape Dimensions	Enlarged How Many Times?	Similar Shape Dimensions
	Length: _____ Width: _____ Perimeter: _____		Length: _____ Width: _____ Perimeter: _____
	Length: _____ Width: _____ Perimeter: _____		Length: _____ Width: _____ Perimeter: _____

Practice: 2 Times, 5 Times, 10 Times

List what the dimensions would be for each shape shown below if it were enlarged 2 times, 5 times, and 10 times.

1.

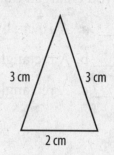

3 cm   3 cm

2 cm

2 times: _____ _____ _____

5 times: _____ _____ _____

10 times: _____ _____ _____

2.

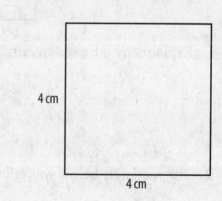

4 cm

4 cm

2 times: _____ _____

5 times: _____ _____

10 times: _____ _____

3.

1 cm

5 cm

2 times: _____ _____

5 times: _____ _____

10 times: _____ _____

Practice: Similar? True or False?

Look at the diagram and read the statements. Mark each statement "True" or "False." If any statement is false, explain why.

1. If you enlarged this shape 5 times, the new shape would be

    __T__ **a.** A 10 cm by 20 cm rectangle.

    __x__ **b.** A rectangle with a perimeter of 30 cm.

    __x__ **c.** A rectangle with a length of 2 cm.

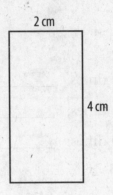

2 cm

4 cm

    Explain any false statements.

2. The rectangle below was enlarged to 10 cm by 25 cm.

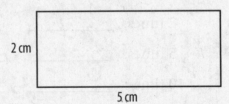

2 cm

5 cm

    __T__ **a.** The original perimeter measured 14 cm.

    __T__ **b.** The enlarged perimeter measured 70 cm.

    __T__ **c.** The original shape was enlarged five times to create the new one.

    Explain any false statements.

3. This shape was enlarged from an original.

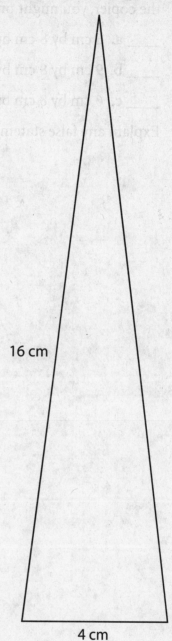

16 cm

4 cm

The original shape could have had

___ **a.** A perimeter that measured 80 cm.

___ **b.** A side that measured 8 cm.

___ **c.** A side that measured 2 cm.

Explain any false statements.

4. An original triangle measures 3 cm by 4 cm by 5 cm. Playing with the copier, you might produce enlargements that measure

_____ **a.** 6 cm by 8 cm by 15 cm.

_____ **b.** 9 cm by 8 cm by 15 cm.

_____ **c.** 6 cm by 8 cm by 12 cm.

Explain any false statements.

Practice: Perimeter Problems

Perimeter problems often involve situations like fencing a space, installing baseboard, or measuring the distance around a field or park.

1. List three situations in which you can see a perimeter.

   a. _____

   b. _____

   c. _____

2. Sketch each situation in a way that highlights the perimeter.

   a.

   b.

   c.

3. Write a perimeter word problem using one of these situations.

Practice: Similar or Not?

Which of the pairs show similar shapes?

Use a ruler to measure and prove your answer.

Put an "S" on all similar shapes and write the measurements next to the sides.

1.

A.

B.

2.

A.

B.

3.

A.   **B.**

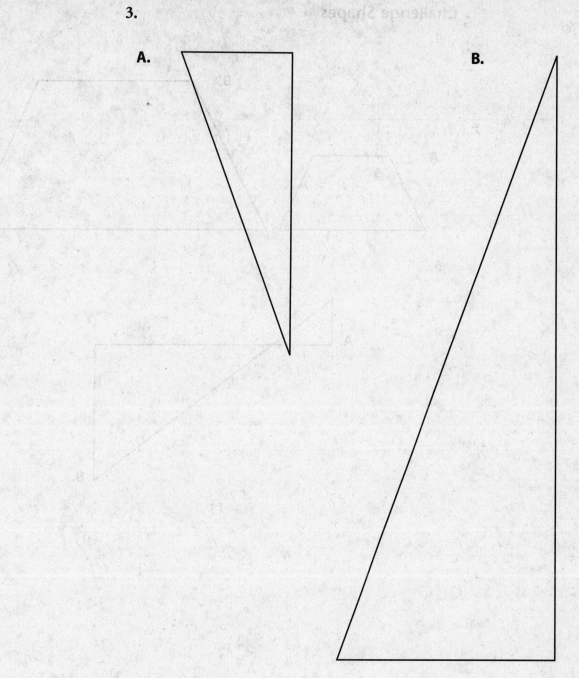

Challenge Shapes

4.

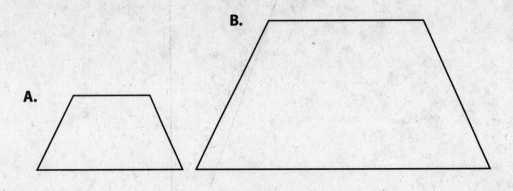

5.

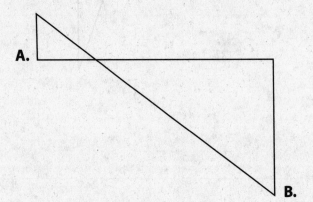

Test Practice

1. Triangle A is similar to Triangle B; the sides of Triangle A are twice as long as those of Triangle B. What is true about the angles in these two triangles?

    (1) The angles' measures are the same.

    (2) They must include a 90° angle.

    (3) Together the six angles add up to 180°.

    (4) Divide 180° by three to find the measure of each angle.

    (5) Triangle A's angles are twice as big as Triangle B's.

Questions 2 and 3 refer to the shapes drawn below.

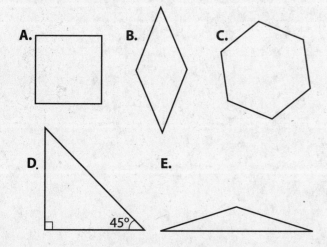

2. Julie drew five shapes as shown above. Which of Julie's shapes have at least one angle greater than 90°?

    (1) Shapes A, B, and C

    (2) Shapes B, C, and E

    (3) Shapes B, D, and E

    (4) Shapes C, D, and E

    (5) Shapes A and D

3. Which of Julie's shapes have at least one angle that measures less than 90°?

    (1) Shapes B, C, and E

    (2) Shapes A and D

    (3) Shapes B, D, and E

    (4) Shapes C and E

    (5) Shapes B and C

4. Sam wants to draw a triangle similar to, but not exactly like, Dan's. Dan drew a triangle that measured 3 cm by 4 cm by 5 cm. Which of the following could represent the dimensions of a triangle similar to Dan's?

    (1) 4 cm by 5 cm by 6 cm

    (2) 5 cm by 6 cm by 7 cm

    (3) 6 cm by 8 cm by 9 cm

    (4) 6 cm by 8 cm by 10 cm

    (5) 12 cm by 13 cm by 14 cm

5. Jane cut out a rectangle for her quilt and asked Lisa to cut out a similar one that was three times as big. How will Lisa know what dimensions her rectangle should be?

    (1) She will multiply the length by 3″.

    (2) She will add 3″ to the length.

    (3) She will multiply the width by 3″.

    (4) She will add 3″ to the perimeter.

    (5) She will multiply the length by 3″ and the width by 3″.

6. What is the maximum number of right angles that can be drawn within a straight angle?

LESSON 5

Line Up by Size

> *How do the sizes compare?*

Is it bigger or smaller? Many times we want to make comparisons. But what does "bigger" or "smaller" mean? Two pairs of pants could be the same length, but one might have a larger waist. In this lesson, you will have a chance to work with the puzzle of which is bigger.

You will explore *what* to measure *when*. In real-life situations, you might need to measure the edges (perimeter) if you are building a fence, or the **area** if you are buying carpet. Sometimes people have a hard time understanding the difference, but it is important to know exactly what kind of measurement you need to make.

Activity 1: Line Up Shapes by Area

Work through the problems with your group.

"Eyeball" It First

- On your own, arrange Shapes 1–6 by *area* size from smallest to largest. Make quick "eyeball" decisions.

- Post your order in the chart on the next page.

- Check those shapes that are *not* in the same order as your partner's.

Compare Areas Second

- This time arrange shapes by area size from smallest to largest with more care.

- *Hint:* Compare sizes as you did with the shoes at the start of class, or use another method.

- Check with your partner to see if you both agree.

- Check with your whole class to see if you all agree.

Look for Relationships

- Now that you have the shapes ordered by size, look at the smallest shape. How does that compare to the one that is next in order?

- Look at the largest shape. How does it compare to the next largest one?

Activity 2: Line Up Shapes by Perimeter

Once the class has agreed on and discussed the ordering of shapes by area size, do the following:

Measure Around

1. "Eyeball" the perimeter of the shapes.

2. Use your meter sticks or centimeter rulers to find the perimeter of the shapes.

3. In the chart, list the shapes from smallest to largest by perimeter.

4. For each shape, write down the shape number, the length of each side (the **dimensions**), and the perimeter.

Arrange the Shapes	Smallest to Largest Your Order Here
Quickly "eyeball" the area	
Compare areas with care	
Compare perimeters with care	

Practice: Seeing Perimeter and Area

Use a colored marker, pencil, or pen to show the perimeter of each shape below. Use a different color to show the area.

EMPower™

Practice: The Job Demands...

1. Next to each job, write what measurements the job demands. You may use more than one letter.

    A for area

    P for perimeter

    D for dimensions (length or width)

    a. Buy wall-to-wall carpet   _____ 3

    b. Put seed down for grass   _____ 1

    c. Buy tiles for the roof   _____ 1

    d. Buy stone for a walkway   _____ 1

    e. Plant trees along a fence   _____ 2

    f. Buy an air conditioner for a room   _____ 1

    g. Find the length of a jogging path around a pond   _____ 2

    h. Compare the land size of two islands   _____ 3, 1

    i. Buy tape to put around a window   _____ 2

    j. Do the floor plan for a house   _____ 2

    k. Plan a ride along a river   _____ 3, 2

    l. Buy a tablecloth   _____ 1

    m. Install a sidewalk   _____ 1

2. Pick two jobs from the list above and explain why you chose A, P, and/or D.

3. What are two other situations in your life that involve figuring out area?

4. What are two other situations in your life that involve figuring out perimeter?

Practice: The Caterers' Question

Two friends, Rachel and Mike, baked cakes for a
party they catered. They made the same rectangular
cake for each table of six. When they cut the cakes,
they decided to be creative. Their goal was to cut each cake into six
equal-size pieces, but different shapes.

Rachel cut the cake for one table like this:

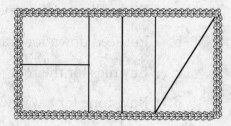

Mike cut the cake for another table like this:

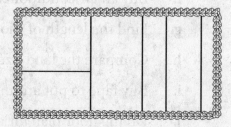

1. At each table, a person complained, "My piece looks smaller than
   hers. That is not fair." Would you agree? Explain.

2. Make your own creative cuts in the two cakes below so that the six
   people each get the same size piece. Show how you know they are
   equal.

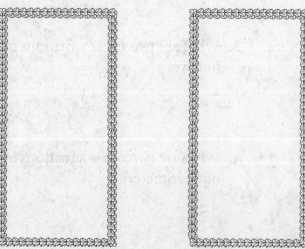

Extension: Area and Perimeter in My Neighborhood

Look around your neighborhood. Pay attention to the different shapes you see.

1. Find something that reminds you of perimeter.

   a. What is it?

   b. Describe or draw it.

   c. What about it reminds you of perimeter?

2. Find something that reminds you of area.

   a. What is it?

   b. Describe or draw it.

   c. What about it reminds you of area?

3. Was it easier for you to find perimeter or area? Why?

1. Which shape has the largest perimeter?

   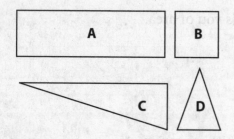

   (1) A is largest.

   (2) B is largest.

   (3) They are all equal.

   (4) C is largest.

   (5) There is insufficient information to decide.

2. For which two shapes could you say that the area of one is half the area of the other?

   (1) A and B

   (2) A and C

   (3) B and C

   (4) B and D

   (5) A and D

3. Erin plans to make a dog pen. She has on hand 24 feet of fencing. If she makes the pen a rectangle, which of the following could work as the dimensions? (l = length and w = width)

   (1) $l = 10, w = 2$

   (2) $l = 6, w = 6$

   (3) $l = 8, w = 4$

   (4) All of the above

   (5) None of the above

4. Which of the following is an example of a perimeter?

   (1) The amount of water for a pool

   (2) The length of a water hose

   (3) The height of a building

   (4) The length of lace trimming on a tablecloth

   (5) The amount of writing space on a page

5. Which of the following is an example of area?

   (1) The amount of space available on a countertop

   (2) The length of a suitcase

   (3) The width of a patio

   (4) The depth of pipe needed for drainage

   (5) The edging around a garden pond

6. Find the shape with the largest perimeter. What is the measure of the perimeter?

   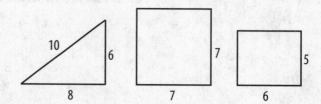

Combining Rectangles

> *How much area does the hopscotch game take?*

In this lesson, you will draw rectangles and then combine them to form a new, composite shape. When you compare measurements of the original rectangles with those of your new shape, you will see how combining rectangles affects area and perimeter.

Rectangles are among the most basic shapes in the world. You see them, for example, in boxes, buildings, street blocks, photographs, and magazines. When you know how to measure rectangles, you can use that information to make things, as well as to solve problems.

You cannot measure area with a ruler because rulers measure length in *lines*. Area is measured in *square* units. How many **square centimeters** (sq. cm) are inside this shape?

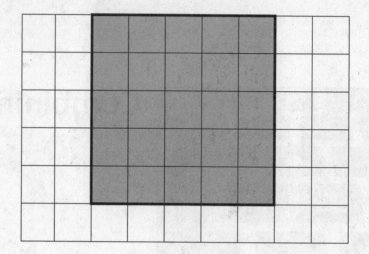

Activity 1: Drawing Four Rectangles

- Draw a 5 cm x 10 cm rectangle on square-centimeter grid paper. Label it "Rectangle 1." Draw three more rectangles of different dimensions on the grid paper. Label them "Rectangles 2, 3, and 4."

- Record the measurements in the table below. Use *cm* for length, width, and perimeter measures and *sq. cm* for area.

	Length (*l*) in cm	Width (*w*) in cm	Area (*A*) in sq. cm	Perimeter (*P*) in cm
Rectangle 1				
Rectangle 2				
Rectangle 3				
Rectangle 4				
All Rectangles Combined				

After you are done, ask a partner to check your measurements. Make sure you both agree the information you each recorded is accurate.

Activity 2: Making a Composite Shape

Imagine cutting out the four rectangles and then arranging them to make one combined shape.

1. Do you predict the area of the new combined shape will be larger, smaller, or the same as the total area of the four shapes you started with?

2. Do you predict the perimeter of the new combined shape will be larger, smaller, or the same as the sum of the perimeters of the four rectangles you started with?

Now actually cut out your four rectangles and tape them together carefully to make one new combined shape. *No overlaps, no gaps!*

3. What is the area of the new shape? How do you know?

4. What is the perimeter of the new shape? How do you know?

5. How many sides does the shape have?

6. How many angles does the shape have? Are they all 90° angles?

My Composite Shape (a sketch)	Area (*A*) sq. cm	Perimeter (*P*) cm	Number of Angles	Number of Sides

Practice: Area of 24 Sq. Cm

1. How many different rectangles can you design that have an area of 24 square centimeters?

   • Use square-centimeter grid paper to draw at least five rectangles that have an area of 24 sq. cm.

   • Find the perimeter of each of these shapes.

   • Complete the chart below.

2. Which rectangle has the smallest perimeter? _____
   Describe its shape.

3. Which rectangle has the largest perimeter? _____
   Describe its shape.

Rectangle Dimensions (length and width)	Perimeter of Rectangle (Show or explain work.)	Area of Rectangle
A. l = w =		24 sq. cm
B.		
C.		
D.		
E.		

Practice: Divide the Shapes

Add lines to the shapes below to show how to make finding the area easy.

1.

2.

3.

4.

5.

6.

7.

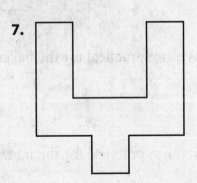

8.

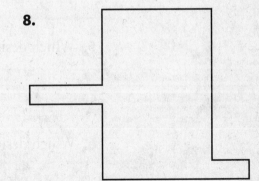

Practice: Cookie Cutter

You are starting a bakery and want to have unique cookie shapes.

1. Use grid paper to design two shapes for cookie cutters. All lines should be straight (no curves). Both shapes have the same area, 36 sq. cm. When you finish, measure the sides and find the perimeters for both cookie cutters.

   Cookie Cutter 1

   Area: _____

   Perimeter: _____

   Cookie Cutter 2

   Area: _____

   Perimeter: _____

2. You are ready to submit your designs to a metalsmith who will make the cookie cutter. Which number should you provide—perimeter or area? Why?

3. The sales person tells you that the cookie cutters will cost 10¢ for every centimeter. Which cookie cutter design is cheaper?

4. Which design would you choose? Why?

5. Which design is more practical for the baker? Why?

6. Which design is more practical for the metalsmith? Why?

Practice: Area in Packaging

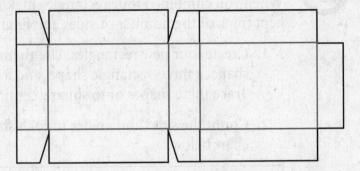

- Find a small box, for example, a box for butter or raisins.

- Open up the box and lay it flat.

- Trace the box on grid paper.

1. About how many square centimeters of cardboard were needed to construct the box?

2. Make a sketch and explain how you arrived at the answer. You may use grid paper.

> Draw lines to make rectangles within the box shape; label dimensions.

Extension: Sides and Angles

When you combined four rectangles making a **composite shape**, you kept track of the number of sides and angles in the shape.

1. Create four new rectangles. Use them to make a two-rectangle shape, a three-rectangle shape, and a new four-rectangle shape. Trace these shapes onto square-centimeter grid paper.

2. Count the sides and angles in each shape. Record your data in the chart below.

Number of Rectangles	Number of Sides	Number of Angles
2 Rectangles		
3 Rectangles		
4 Rectangles		

3. Look at the numbers you recorded for sides and angles. What do you notice?

4. Does this surprise you? Why?

5. Are all the angles right angles? Why do you think this is so?

6. If a shape had eight angles, how many sides would you expect it to have? Why?

> Think about putting eight angle demonstrators together to form a shape made of rectangles. What would happen?

1. Which of the following is *always* true about a rectangle?

   (1) It has four angles that total 180°.

   (2) It has 4 right angles and 4 sides.

   (3) It has only 2 right angles and 4 sides.

   (4) It has 4 right angles and 4 sides with equal widths.

   (5) It has only 2 right angles and 4 sides with opposite sides equal.

2. Which of the following is *not* a situation involving area?

   (1) Finding the amount of material needed to cover a pool

   (2) Finding the space in a parking lot

   (3) Finding the size of a lawn

   (4) Finding the length of wood needed for a picture frame

   (5) Finding how much of the Earth is covered in water

3. A rectangle's area measures 12 square centimeters. Which of the following might be the dimensions for the rectangle.

   (1) $l = 6; w = 3$

   (2) $l = 12; w = 2$

   (3) $l = 6; w = 2$

   (4) $l = 4; w = 2$

   (5) $l = 4; w = 4$

4. Candi and Claude each have a piece of chocolate. Candi's chocolate piece measures 6 cm by 4 cm, while Claude's piece measures 5 cm by 4 cm. How much bigger is Candi's chocolate than Claude's?

   (1) 4 cm

   (2) 4 sq. cm

   (3) 5 cm

   (4) 6 cm

   (5) 6 sq. cm

5. Sarah cuts out three rectangles. She finds the area and perimeter for each one. Sarah adds the areas of all the rectangles and adds all the perimeters. She then pastes the rectangles together (picture below). When she pastes them together into a new shape, what is true about the new shape?

6. Find the perimeter of Sarah's composite shape.

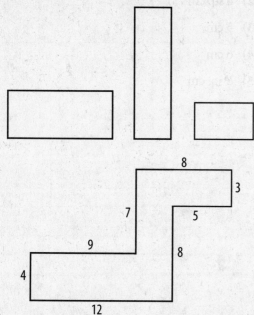

(1) The number of sides is the same as before.

(2) The number of angles is the same as the number of sides.

(3) The area is larger than before.

(4) The perimeter is the same as before.

(5) The area is smaller than before.

Disappearing Grid Lines

> **What is the missing information?**

At times you have only a few facts about a rectangle. You might know the dimension of one side and the total area, and need to find the perimeter; or you might have the dimension of one side and the perimeter, and need to find the area. In this lesson, you will work with the rules that will help you find the missing information in these situations.

Activity 1: Missing Measurements

The table below shows some information about eight rectangles.

1. Draw the rectangles on square-centimeter grid paper.

2. Fill in the chart with the missing measurements.

Shape	Length (*l*)	Width (*w*)	Area (*A*)	Perimeter (*P*)
1	10 cm	6 cm		32 cm
2	12 cm	5 cm		
3	4 cm		12 sq. cm	
4	4 cm			12 cm
5		4 cm		40 cm
6	12 cm		24 sq. cm	
7			144 sq. cm	48 cm
8			144 sq. cm	52 cm

3. How did you find the missing information?

Activity 2: When the Grid Lines Disappear

1. Fill in the chart with the missing information.

Shape	Length (*l*)	Width (*w*)	Area (*A*)	Perimeter (*P*)
1. 3 cm / 5 cm				
2. 8 cm / 10 cm				
3. A = 100 sq. cm P = 58 cm				
4. 40 cm P = 160 cm				

Activity 3: Area of Right Triangles

1. Use the triangles your teacher gives you and the square-centimeter grid paper below to find the area of each triangle.

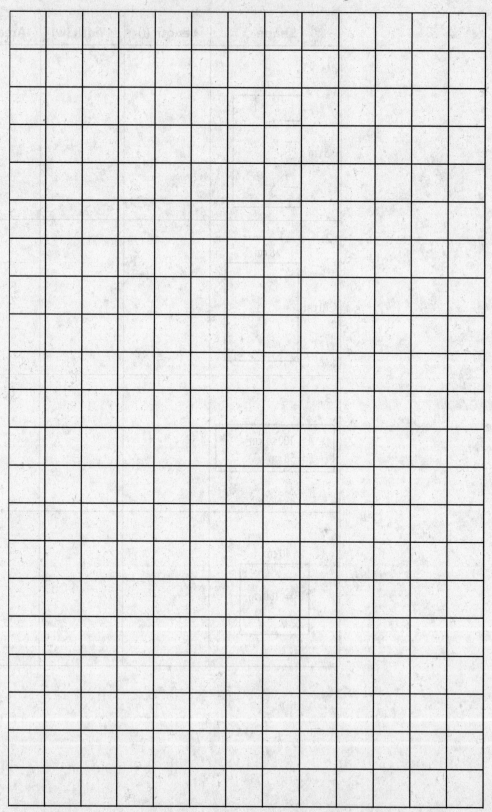

Activity 3: Area of Right Triangles *(continued)*

Knowing how to find the area of rectangles can help you find the area of right triangles. Remember the square and triangle from the shape set?

2. How many triangles fit inside the square? _____

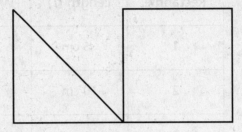

3. You can find the area of the *square* by using the following formula:

_____ = Area

4. What can you do to find the area of one triangle?

To see how a triangle's area and a rectangle's area are related, first create a solid rectangle by adding on another triangle. Here is one example:

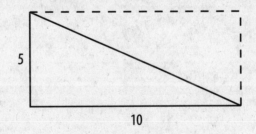

Area of the rectangle = 50

Area of the triangle = 25

Practice: Missing Measurements

1. Draw the rectangles on square-centimeter grid paper.

2. Fill in the chart with the missing measurements.

Rectangle	Length (*l*)	Width (*w*)	Area (*A*)	Perimeter (*P*)
1	5 cm	4 cm		18 cm
2	10 cm	3 cm		
3	6 cm		24 sq. cm	
4		3 cm		16 cm
5		20 cm		100 cm
6	15 Cm	2 cm	30 sq. cm	
7	60	3	180 sq. cm	184 cm
8	60	3	180 sq. cm	98 cm

3. How did you find the missing information?

Practice: More Complex Shapes

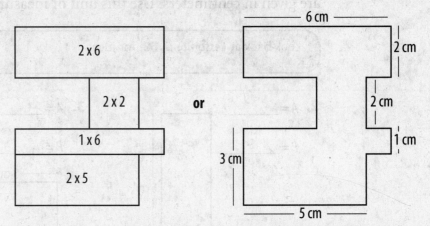

1. The figure above on the left shows the dimensions for four rectangles. The figure on the right shows the composite shape of those rectangles.

   a. What is the sum of the areas of the four rectangles?

   b. What is the area of the composite shape?

   c. What is the sum of the perimeters of the four rectangles?

   d. What is the perimeter of the composite shape?

Find the combined area and perimeter of Shapes 2–5. All measurements are given in centimeters. Use this unit of measurement in your answer.

> Look back at *Lesson 6*, p. 63, for ideas.

2. $A =$ _____

$P =$ _____

6 cm

2 cm

2 cm

6 cm

3. $A =$ _____

$P =$ _____

3 cm

2 cm

2 cm

6 cm

4. $A =$ _____

$P =$ _____

5 cm

3 cm

2 cm

8 cm

20 cm

5. $A =$ _____

$P =$ _____

5 cm

15 cm

25 cm

15 cm

20 cm

Practice: Von's Kitchen

Von wants to tile his kitchen. The floor is a rectangle 12 feet long and 10 feet wide. How big is his kitchen?

Make a sketch of Von's kitchen floor. Label all dimensions. Answer the following questions, and show all your work.

1. Area = _____

2. Perimeter =_____

3. The method I used to find area:

4. The method I used to check area:

5. My *preferred* method and why I like it:

6. The method I used to find perimeter:

7. The method I used to check perimeter:

8. My *preferred* method and why I like it:

Practice: Areas of More Right Triangles

Outline the shape of the rectangle that could be created with two of the same triangles for each of the problems below.

Then find the area of each triangle and the area of the rectangle that could be created with two of the same triangles. The first problem has been started for you.

1. Area of the rectangle: _____

Area of the triangle: _____

2. Area of the rectangle: _____

Area of the triangle: _____

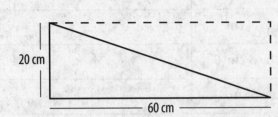

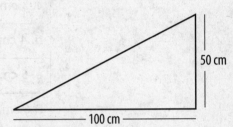

3. Area of the rectangle: _____

Area of the triangle: _____

4. Area of the rectangle: _____

Area of the triangle: _____

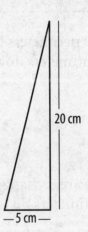

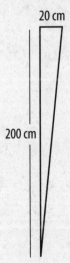

Extension: Area and Perimeter Challenges

1. On grid paper, draw 10 squares: a 1 cm by 1 cm square, a 2 cm by 2 cm square, and so on, up through a 10 cm by 10 cm square. Then fill in the table below.

Square Dimensions	Perimeter	Area
a. 1 cm by 1 cm		
b. 2 cm by 2 cm		
c. 3 cm by 3 cm		
d. 4 cm by 4 cm		
e. 5 cm by 5 cm		
f. 6 cm by 6 cm		
g. 7 cm by 7 cm		
h. 8 cm by 8 cm		
i. 9 cm by 9 cm		
j. 10 cm by 10 cm		

2. Describe how the perimeters change from one square to the next. By how many centimeters do the perimeters change each time?

3. Describe how the areas change from one square to the next. By how many square centimeters do the areas change each time?

4. What patterns do you see?

1. A billboard stands 10 feet high and 20 feet wide including a frame 1 foot wide all the way around the inside edge. The area of the entire billboard (including the frame) is

   (1) 200 feet

   (2) 200 square feet

   (3) 1,200 square feet

   (4) 2, 000 square feet

   (5) 20,000 square feet

2. Since the frame takes up space, the actual space for an ad is smaller. About how much space is available for an ad?

   (1) 55 sq. ft.

   (2) 100 sq. ft.

   (3) 144 sq. ft.

   (4) 170 sq. ft.

   (5) 196 sq. ft.

3. A rug that is 10 feet x 12 feet is how much larger than a 6 foot x 8 foot rug? Which of the following expressions represents the difference in area between a 10 x 12 foot rug and a 6 x 8 foot rug?

   (1) $(10 \times 12) - (6 \times 8)$

   (2) $(10 + 6) \times (12 + 8)$

   (3) $(10 - 6) \times (12 - 8)$

   (4) $(12 \times 10) \div (8 \times 6)$

   (5) $(12 - 10) \times (8 - 6)$

4. Mai is remodeling her 10 foot by 10 foot den. First she builds a one-foot deep bookcase across the length of one wall. If Mai wants to tile the room's floor, she should buy enough tiles to cover what amount of space?

   (1) 19 square feet

   (2) 20 square feet

   (3) 90 square feet

   (4) 99 square feet

   (5) 100 square feet

5. Kate wants to find the area of a triangle-shaped fabric scrap. She knows a right triangle can be seen as half a rectangle. Hers is a right triangle with one side of 90 cm and a base of 40 cm. What is the area of the triangle?

   (1) 320 sq. cm

   (2) 400 sq. cm

   (3) 900 sq. cm

   (4) 1,800 sq. cm

   (5) 3,600 sq. cm

6. Gail originally had a garden plot that measured 30 feet by 40 feet. She added another 20 foot by 20 foot section. What is the area of her new garden in square feet?

Conversion Experiences

When would you measure in inches? In feet? In yards?

In this lesson, you will list items measured in inches, feet, and yards. You will measure the perimeter of some rectangles and compare your results with those of other students. You will also **convert** by changing one unit of measure to another, for example, inches to yards.

Activity: It's All in How You Measure

In your classroom, you will see three rectangles outlined in masking tape. Your group will find the perimeter for each of these rectangles in the unit of measure your teacher assigns.

Group (letter and names of members):_____

_____

Assigned unit of measure:_____

1. List the dimensions and the perimeter of each rectangle below.

Rectangle	Dimensions (*l* and *w*)	Perimeter (*P*)
1		
2		
3		

2. Describe how you found the perimeter for each rectangle. Tell what tool(s) you used and how you calculated the perimeter.

3. Copy the perimeter measurements from the class chart for each of the three rectangles in inches, feet, and yards, and fill in the table below.

Rectangle	Perimeter Measures		
	Inches	Feet	Yards
1			
2			
3			

4. Are all of the perimeter measurements for Rectangle 1 equal? How do you know?

5. Are all the perimeter measurements for Rectangle 2 equal? How do you know?

6. Are all the perimeter measurements for Rectangle 3 equal? How do you know?

7. Choose one dimension of a wall or floor to measure. Present findings in all three units of measure—inches, feet, and yards. Be prepared to respond to these questions:

   a. What did you measure?

   b. What is the length in inches, feet, and yards?

   c. How did you find inches, feet, and yards?

8. Use what you have learned about converting linear units to fill in the blanks:

a. 120″ = _____ feet

b. 36′ = _____ yards

c. 72″ = _____ yards

d. 84″ = _____ feet

e. 180′ = _____ yards

f. 108″ = _____ feet

Practice: Home Measurements

- Choose five objects around your home to measure. Some ideas are doors, counter tops, toaster, oven, and refrigerator.

- Estimate the length in inches, centimeters, or feet.

- Make an actual measurement rounding to the nearest whole unit.

Object	Unit Measurement	Estimated Measurement of Length	Actual Measurement of Length
1.	Inches (in. or ″)		
2.	Feet (ft. or ′)		
3.	Inches (in. or ″)		
4.	Centimeters (cm)		
5.	Your Choice _____		

Practice: Units of Measure

Fill in the chart below:

Linear Unit	Abbreviation	Body Part or Object (to help remember estimated size)
Inches		
Feet		
Yards		
Centimeters		
Meters		

Practice: Keeping Units Straight

1. What are the length and width of this page?

   a. In inches?

   b. In centimeters?

2. What is the perimeter of this page?

   a. In inches?

   b. In centimeters?

3. Fill in the blanks. A wavy equal sign (≈) means "about."

   a. 1 inch ≈ _____ cm

   b. 6 inches ≈ _____ cm

c. 1 foot = _____ inches

d. 3 feet = _____ inches

e. 6 feet = _____ inches ≈ _____ cm

4. What is your height?

   a. In feet and inches?

   b. In inches?

   c. In centimeters?

Extension: Conversion Match

1. When units of measure are converted, they are changed from one type of unit into another. Match each unit of measure on the left with its equivalent measure in another type of unit on the right.

Unit of Measure	Equivalent Unit(s)
a. ___ 1 yard (yd.)	**(1)** 60 minutes (min.)
b. ___ 1 mile (mi.)	**(2)** 12″
c. ___ 1 quart (qt.)	**(3)** 2 pints (pt.)
d. ___ 1 pound (lb.)	**(4)** 2,000 pounds
e. ___ 1 hour (hr.)	**(5)** 5,280′
f. ___ 1 minute (min.)	**(6)** 16 ounces (oz.)
g. ___ 1 ton	**(7)** 32 fluid ounces (fl. oz.)
h. ___ 1 acre	**(8)** 60 seconds
i. ___ 1 foot (ft.)	**(9)** 43,560 square feet
	(10) 36″

Each unit listed, except one, has one equivalent that matches it. One of the units on the left has two equivalents on the right. Which one?

2. For this problem, refer to the left column, "Unit of Measure," in Problem 1. Determine what three of each amount would equal. Then figure out what one half of each amount would equal. Use your calculator if necessary, and record your answers in the chart below.

Unit of Measure and Its Equivalent	Triple the Amount	Half the Amount
a. 1 yd. = 36 in.	3 yd. = 108 in.	$\frac{1}{2}$ yd. = 18
b. 1 mi. =		
c. 1 qt. =		
d. 1 lb. = 16	48	8
e. 1 hr. = 60	180	30
f. 1 min. = 60	180	30
g. 1 ton = 2,000	6000	1000
h. 1 acre = 43,000		
i. 1 ft. = 12	36 inc	6

Extension: The Foreman's Problem

Three co-workers were supposed to measure the foundation sills for a new building. Each had his own tape for measuring. They took their measurements, then wrote them on the foreman's sketch of the project. Below is the sketch with the measurements listed, just as the foreman received it.

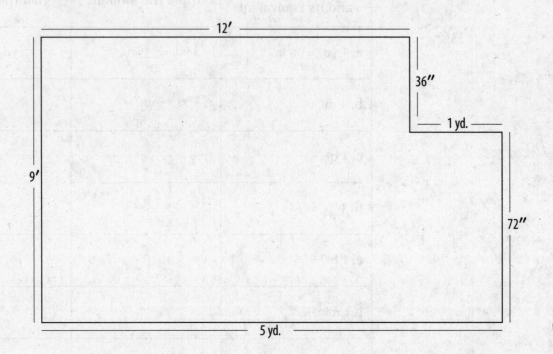

1. Does it appear that the co-workers' measurements are accurate? How do you know?

2. The foreman is supposed to order lumber to place along the foundation perimeter. How much lumber does she need to order? How do you know?

3. If the lumber she buys costs 90 cents per foot, how much will the lumber for this job cost?

Test Practice

1. A rug shop sells runners of different lengths: 18″, 2′, 6′, 45″, and 90″. If Zach has a hall 10 feet long, which combination of runners would come closest to covering the length of his hall?

    (1) 18″ and 2′

    (2) 45″ and 45″

    (3) 90″ and 6′

    (4) 6′ and 45″

    (5) 2′ and 6′

2. Three friends call about a recent snowstorm. Ana reports that 22″ of snow fell at her place, Tex reports that 2′ fell at his, and Lin reports that $\frac{1}{2}$ yd. fell at hers. Which of the following shows the amount of snowfall at the friends' places in order from greatest to least.

    (1) Lin, Tex, Ana

    (2) Ana, Lin, Tex

    (3) Tex, Lin, Ana

    (4) Tex, Ana, Lin

    (5) Ana, Tex, Lin

3. A football field measures 100 yards by 30 yards (not including the seating area). The fencing comes in 10-yard rolls. How many rolls are needed to enclose the field?

    (1) 26

    (2) 108

    (3) 260

    (4) 1,080

    (5) 2,600

4. The window in Charlie's home office measures 3 feet by 6 feet. What is the perimeter of the window?

    (1) 9 feet

    (2) 9 square feet

    (3) 15 feet

    (4) 18 feet

    (5) 18 square feet

5. The maximum rise (height) for a step is 8″. If you want to build a concrete stairway into a 4-foot-deep pool, how many risers do you need?

    (1) Two

    (2) Four

    (3) Six

    (4) Eight

    (5) Ten

6. Jonathan's garden plot is one yard wide and 5 feet long. What is the perimeter, in inches, of his garden plot?

9

Squarely in English

How is this sold: by the square inch, square foot, or square yard?

In the previous lesson, you learned how to convert measurements for length, width, and perimeter. In this lesson, you will explore the square units used to measure area, such as a **square inch**, a **square foot**, and a **square yard**. You need to know area when you buy many things, such as carpet, gravel, or paint because you need to know how much surface to cover. When you know how to convert between square measurements, you can be a more savvy consumer.

Activity: Squarely in English

Group 1: Construct at least 12 square inches.

1. How many square inches are in a square foot? _____

2. Measure the area of one object in square inches.

   Object: _____   Area = _____

3. Sketch the results to share with the group.

4. Recommend a way to remember the size of an inch and a square inch.

Group 2: Construct at least 10 square feet.

1. How many square feet are in a square yard? _____

2. Measure the area of one object in square feet.

   Object: _____   Area = _____

3. Sketch the results to share with the group.

4. Recommend a way to remember the size of a foot and a square foot.

Group 3: Construct at least three square yards.

1. How many square inches are in a square yard? _____

2. Measure the area of one object in square yards.

   Object: _____   Area = _____

3. Sketch the results to share with the group.

4. Recommend a way to remember the size of a yard and a square yard.

Practice: Choose a Unit

1. Match a unit with each space. You can use any unit more than once. Write the letter or letters of the unit on the lines next to each description on the right.

a. Square Inches

_____ the length of the window frame

_____ the surface space of a window sill

b. Square Feet

_____ the space the stamp takes up on an envelope

_____ the length of two tables pushed together

c. Square Yards

_____ metal to construct a street sign

_____ floor space of a child's playpen

d. Inches

_____ size of a calendar page

_____ tiles to cover a hotel lobby floor

e. Feet

_____ height of a swing

_____ length of a pool

_____ distance to the nearest bathroom

f. Yards

_____ victory banner covering an entire wall of a school

_____ depth of a bookshelf

_____ aluminum foil to cover a baking pan

2. Choose one of your answers and explain your reasoning.

Practice: Roll Out the Carpet

1. In this ad, is the price of a square foot of carpet more expensive, less expensive, or the same as the price per square yard? Explain why you think so.

2. How many square inches are in one square foot of carpet? Make a sketch.

3. Some people say it is a good idea to buy a little extra material any time you do a home improvement project. Tony, a carpet salesperson, suggested buying an extra two square yards.

   a. How much extra would that cost?

   b. Does Tony's recommendation make sense to you? Why?

4. There are 1,760 yards in a mile. How many square yards are in a square mile? How do you know?

5. To welcome the visiting president, a town wanted to lay down a mile of red carpet four feet wide. What would such a carpet cost?

Practice: The Better Deal

Moe and Vic bought an older home with a broken 3′ x 3′ window they want to replace when they can afford a new one. For the time being, though, they have decided to cover up the window.

1. They have two choices for covering the window. They can use either plywood that sells for about 45 cents per *square foot* or 36″-wide space-blanket material that sells for about $4.50 per *square yard*. Both materials are equally protective and easy to install. Which is more cost effective? Show your steps.

2. How many square feet of glass would Moe and Vic have left over if they cut a new window from a piece of plate glass that measured 2 yards by 2 yards? Show all work.

3. If they decide to trim the new window with 3″ strapping, how much would they have to buy? *Hint*: Draw the window and the frame.

Practice: San Diego Construction Company

- Fill in the blanks using the numbers and words from the box below so this story makes sense. Use each word or number once. You may use a calculator.

- Sketch the park, the sidewalk, and fencing and label all dimensions; it will be helpful.

The San Diego Construction Company started constructing a fence around a park. They marked the boundaries of the park with 210 feet of tape. Before ordering the fencing at $6.75 per foot, two workers checked the measurements. Since the park was shaped like a _____, they only needed to measure the_____ and the _____. The length was 75 feet. The 210-foot figure was correct, so the width measured _____. At $6.75 per foot, the fencing would cost about $_____.

The plan called for a tiled walkway straight across the center of the park, parallel to the length. The walkway's dimensions were 75 feet long and 4 feet wide. The company planned to lay _____ one-sq.-ft. cement tiles. At $2.59 per tile, the estimated cost was $_____.

300	rectangle	780	width
	1,400	30	length

Practice: Area Measurements

1. Find the area of *one* of the following objects: book cover, calendar, picture, window, or table.

   a. Name the object: _____

   b. Measure the dimensions of the object.

   c. Sketch the object. Label the drawing with dimensions.

   d. Find the area of the object. $A =$ _____

2. Find the area of each of the following rectangles.

Rectangle	Dimensions	Area	Steps you took to find area of Rectangle 1
1	18 in. x 3 ft.		
2	3 ft. x 6 ft.		
3	6 ft. x 3 yd.		

3. Sketch a composite shape using the three rectangles.

4. Find the area and perimeter of the composite shape.

Extension: Checkerboard Paint Job

Sue wants to paint a giant checkerboard on her playroom wall. A checkerboard is square itself and contains 64 smaller squares. She wants each square within the checkerboard to measure 9 inches on each side.

1. Find the area of each square within the checkerboard.

2. Find the area of Sue's giant checkerboard in square *feet*.

> A labeled diagram will help with this problem.

Test Practice

1. What could you do to find the number of square *yards* needed for a carpet on a floor that measures 156 square *feet*?

    (1) Divide by 144.

    (2) Divide by 3.

    (3) Multiply by 144.

    (4) Divide by 9.

    (5) Multiply by 9.

2. Which of the following is a true statement describing the difference between an inch and a square inch?

    (1) There are four sides in an inch and in a square inch.

    (2) Inches measure perimeter; square inches measure area.

    (3) Inches are squares, and square inches are lines.

    (4) 12 inches equals one foot; 24 square inches equals one square foot.

    (5) There are 12 feet in an inch and 36 feet in a square inch.

3. What could you do to find the number of square inches in 2 square feet?

    (1) Divide 2 by 144.

    (2) Multiply 2 by 12.

    (3) Divide 2 by 12.

    (4) Multiply 2 by 144.

    (5) Multiply 2 by 48.

4. Which of the following represents units of measurement from smallest to largest?

    (1) Yard, foot, inch, centimeter

    (2) Centimeter, inch, foot, yard

    (3) Inch, foot, centimeter, yard

    (4)) Inch, centimeter, foot, yard

    (5) Centimeter, inch, yard, foot

5. This line is approximately how long?

    _____

    (1) One centimeter

    (2) One foot

    (3) One inch

    (4) One yard

    (5) Three inches

6. The window in Cal's home office measures 3 feet by 6 feet. What is the area of the window in square inches?

10

Scale Down

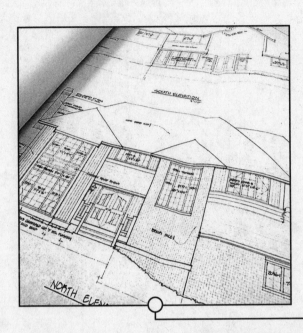

What's the scale ?

In this lesson, you will learn about **scale**. In mathematics scaling allows you to show something large on a much smaller piece of paper. When you scale, you change the size of the original, but all the relationships between dimensions stay the same. You will see how a mathematician makes a scale drawing, and then you will use the same method to make your own scale drawing.

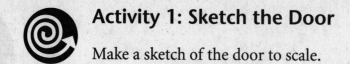

Activity 1: Sketch the Door

Make a sketch of the door to scale.

Activity 2: A Mathematician Uses Scale

Your Job

1. Read below how a mathematician draws a floor plan to scale.

2. Make a list or outline of those steps.

3. Use the same steps to make a scale drawing of a door on a piece of grid paper.

4. Be ready to explain to the class the mathematician's method and why it works.

A Mathematician Speaks

When I do a layout, I use grid paper. First I figure out how many squares are on the paper. I take a measurement in inches, feet, or centimeters. I could choose to have one edge of a square of the grid equal one foot; or if the measurement is in inches, I might have one edge of the square of the grid equal 10 or 20 inches.

When you finish making the first scale drawing of a door, make another. This time use a smaller space on the paper.

1. List your steps:

2. Why does the method work?

Practice: Scale Drawings

1. Refer to your work for *Area Measurements*, page 101, where you found the area of one of the following objects: book, calendar, picture, window, or table. Use grid paper to make a scale drawing of your object. Label the drawing with the scale you used.

2. Choose another object.

    a. Sketch and title it.

    b. Label dimensions with measurements.

    c. Determine the area.

    d. Use grid paper to make a scale drawing of the object. Label the drawing with the scale you used.

Practice: Shrinking Doors

For Problems 1–4, answer the questions, and then make a scale drawing for each door on a separate piece of paper. Include the sign, window, poster, or mail slot in your drawing as well.

1. Bik wants to draw her door to scale. It has a "No Smoking" sign on it that is 1 foot by 1 foot. The door is 7 feet by 4 feet.

   a. What scale could she use? _____ = 1 foot

   b. What would the length and width of the scale drawing of the door be?

      length = _____   width = _____

   c. What are the dimensions of the scale drawing of the sign?

      length = _____   width = _____

2. Bik's basement door measures 6 feet by 3 feet. It has a window that is 2 feet by 1 foot.

   a. What scale could she use? _____

   b. What would the length and width of the scale drawing of the door be?

      length = _____   width = _____

   c. What are the dimensions of the scale drawing of the window?

      length = _____   width = _____

3. Bik's closet door has a poster of three cats on it. The closet door is 6 feet by 5 feet. The cat poster is 18 inches by 24 inches.

   a. What scale could she use? _____

   b. How big would the scale drawing of the door be?

      length = _____   width = _____

   c. How big will the scale drawing of the poster be?

      length = _____   width = _____

4. Bik's front door has a mail slot that is 2 inches tall and 12 inches wide. The door is 84 inches high and 48 inches wide. Use a scale of 1 cm = 12 inches for the following questions:

a. What would the length and width of the scale drawing of the mail slot be? _____

b. What are the dimensions of the scale drawing of the door?

length = _____   width = _____

Practice: All About Shapes

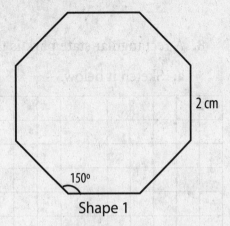

Shape 1

1. A similar shape that is half the height of Shape 1 will have angles that measure _____° and sides that measure _____ cm.

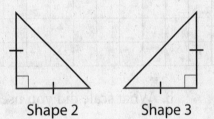

Shape 2   Shape 3

True or False?

2. _____Shapes 2 and 3 can be called isosceles right triangles.

3. _____Shapes 2 and 3 each have one set of parallel sides.

4. _____Shapes 2 and 3 each have one right angle.

5. _____Shape 2 is scaled down from Shape 3.

The two trapezoids below are similar.

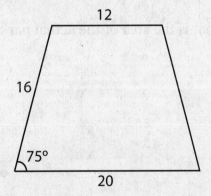

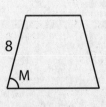

6. What are the missing dimensions of the smaller trapezoid?

7. What is the measure of ∠M? _____

8. A rectangular state park is three miles long and five miles wide.

    a. Sketch it below.

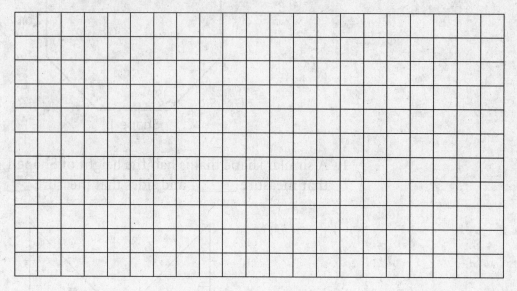

    b. What scale did you use? _____

    c. What is the area of your sketch?

    d. What is the area of the actual park?

Extension: Enlarge It!

Select a photo, card, or drawing.

1. What are its dimensions?

   length = _____   width = _____

Make an enlargement of it on a photocopier. Press "enlarge" or "%" on the copier, and choose a "%" larger than 100.

2. What are the dimensions of your enlargement?

   length = _____   width = _____

3. Compare the dimensions of your original object with those of the enlargement. What is the approximate scale?

Now shrink your original. Photocopy it, choosing a "%" smaller than 100.

4. What are the dimensions of your reduction?

   length = _____   width = _____

5. Compare the dimensions of your original with those of the reduction. What is the scale of the new version?

Test Practice

1. Lena used square-centimeter grid paper to draw her room to scale. Each *foot* was represented by 1 cm. If her room is 12 feet wide, how many centimeters represented the width?

   (1) 2 cm
   (2) 3 cm
   (3) 4 cm
   (4) 6 cm
   (5) 12 cm

2. Lena was not happy with the result of her drawing, so she switched to half-inch grid paper. Then she represented each foot by a half-inch. If her room is 12 feet wide, how many inches will represent the width in her drawing?

   (1) 2 in.
   (2) 3 in.
   (3) 4 in.
   (4) 6 in.
   (5) 12 in.

Questions 3 and 4 refer to a cereal company boss who oversees a new round of ads.

"I want the cereal box larger than life," the boss said. "Make it stand 100 times taller on the billboard."

3. The results for an 8″ x 12″ cereal box would be

   (1) 80′ x 120′
   (2) 800″ x 120″
   (3) 800″ x 12,000″
   (4) 800″ x 1,200″
   (5) 940,000 sq. inches

4. "Too big," said the artist. He was thinking that, in square feet, the box would take up a space close to

   (1) 3,300 sq. ft.
   (2) 6,600 sq. ft.
   (3) 80,000 sq. ft.
   (4) 960,000 sq. ft.
   (5) 1,000,000 sq. ft.

5. Rachel's blueprints are scaled so that 1″ = 3′. On the blueprint, her kitchen measures 4″ wide. Which expression(s) below shows how to determine the number of feet in the width of Rachel's kitchen?

   (1) 1 x 3
   (2) 3 x 4
   (3) 1 x 4
   (4) 4 x 12
   (5) 4 x 1

6. Kevin's map scale says 1″ = 50 miles. When he measures the distance between his home and San Francisco, the distance measures $4\frac{1}{2}$ (4.5) inches. About how far away from San Francisco (in miles) does Kevin live?

Filling the Room

How much will fit in here?

In this lesson, you will build on what you know about measurement and dimensions to figure out how many file boxes would fit inside your classroom.

Have you ever seen an ad for a large-capacity freezer or refrigerator? Sometimes you want to know how many boxes you can stuff into a space like a freezer, closet, or car trunk. It helps to have a way to think about these problems.

Activity: Filling the Room

Pretend your classroom will soon be converted into a storage room. It will be filled with boxes of files because the front-office workers want all their paperwork kept in one spot. Your teacher will show you a sample box.

How many boxes do you think could fit inside the classroom? _____

Make a plan to determine how many boxes can fit into the room. List the steps you will take to solve this problem.

Notes for our plan:

Solution and steps we took to solve the problem:

Practice: How Many Fit?

Kat likes to cook for the whole week on Sunday. She stores her food in 9″ x 12″ x 4″ containers. How many can she stack on one shelf of her refrigerator? The shelf is 30 inches wide, 9 inches high, and 24 inches deep.

Make a sketch and show your work.

Practice: Measuring Things

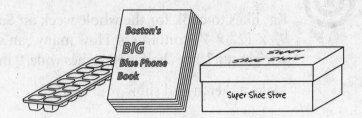

1. With a ruler, yardstick, or tape measure, find the actual length, width, and height of the objects listed.

   Choose a unit for measuring, and circle it.

   Nearest 1 centimeter   Nearest $\frac{1}{2}$ inch   Nearest $\frac{1}{4}$ inch

	length (*l*)	width (*w*)	height (*h*)
a. Phone book	*l* = _____	*w* = _____	*h* = _____
b. Shoe box or jewelry box	*l* = _____	*w* = _____	*h* = _____
c. Ice-cube tray	*l* = _____	*w* = _____	*h* = _____
d. VCR or CD player	*l* = _____	*w* = _____	*h* = _____
e. A door	*l* = _____	*w* = _____	*h* = _____
f. Your choice _____	*l* = _____	*w* = _____	*h* = _____

2. Did anything confuse you when you measured these items? What?

3. Which were easiest to measure? Why?

Practice: Area, Perimeter, or Volume?

Each of these situations involves area, perimeter, or volume. Decide which term applies to the situation. Label each situation *A*, *P*, or *V*.

1. Fencing

   a. Amount of fencing around the local playground: _____

   b. Explain your reasoning. What dimensions would you measure?

2. Trucks

   a. Number of truckloads or trips needed to carry away a pile of dirt: _____

   b. Explain your reasoning. What dimensions would you measure?

3. Paint

   a. Amount of paint you would need to cover the four walls of a hallway: _____

   b. Explain your reasoning. What dimensions would you measure?

4. Wading pool

   a. Amount of water needed to fill a wading pool: _____

   b. Explain your reasoning. What dimensions would you measure?

5. Shingles

   a. Amount of shingles needed to cover the walls of a shed: _____

   b. Explain your reasoning. What dimensions would you measure?

Extension: Pack It In

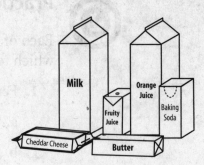

1. Choose one shelf of *your* refrigerator. Describe the shelf in words or pictures. List the dimensions to the nearest inch.

2. Now choose one box or container that you might store in the refrigerator. Describe the box. Use words or pictures. List the dimensions to the nearest inch.

3. How many boxes or containers like the one you described for Problem 2 could fit on your refrigerator shelf? Remember: You can pile them on top of each other.

4. Explain how you found your answer. What steps did you take to figure out the problem?

Test Practice

1. Shana wants to know the volume of her bottom drawer so she can figure out how many packages of paper she can store in there. What must she measure?

   (1) Length only

   (2) Width only

   (3) Height only

   (4) Length, width, and height

   (5) Length and width

2. Phillipe told Alan that he had a huge box that measured 1 yard by $1\frac{1}{2}$ feet by 24 inches. Which of the following represents another way Phillipe might have stated the measurements?

   (1) 1 foot by $\frac{1}{2}$ yard by 24 inches

   (2) 1 yard by 30 inches by 24 inches

   (3) 3 feet by $1\frac{1}{2}$ feet by 2 feet

   (4) 3 feet by $1\frac{1}{2}$ feet by 1 foot

   (5) 36 inches by 30 inches by 24 inches

3. Tom did some measuring.

   Nail boxes: Each one measures 4″ by 4″ by 2″.

   Shelf space: It measures 20″ by 36″ by 8″.

   How many nail boxes will fit on the shelf? (He does not care how the boxes are stacked because they are unopened.)

   (1) 90 boxes

   (2) 180 boxes

   (3) 360 boxes

   (4) 3,600 boxes

   (5) Insufficient information

4. Ms. Stine wants to cover the classroom tables with paper for an art project. There are three tables that measure 2′ by 6′ and two large tables that measure 3′ by 8′. Which expression below shows the square feet of paper Ms. Stine will need to cover all the tables?

   (1) 2′ x 6′ + 3′ x 8′

   (2) 3 x (3′ x 8′) + 2 x (2′ x 6′)

   (3) 5 x (3′ + 8′ x 2′ + 6′)

   (4) 2 x (2′ + 6″) + 2 x (3′ + 8′)

   (5) None of the above

5. A large closet is 10 ft. x 4 ft. x 10 ft. Stacked along one wall are 100 12 in. x 12 in. x 12 in. boxes. Approximately how many of the same size boxes could fit inside the closet?

   (1) 200

   (2) 300

   (3) 400

   (4) 500

   (5) 1,000

6. A grocery store stocker is putting out 24-can cases of cola on a shelf. Each case measures 16″ long by 5″ wide by 11″ high. The shelf is 60″ wide and 12″ deep. How many cases can she fit into one layer on the shelf?

Cheese Cubes, Anyone?

> **How many cubes of cheese will you need for your party?**

When you described area, the units you used were square inches or square feet. In this lesson, you will find volume by measuring length, width, and height, and describe it in cubic units.

You will construct a one-inch cube and solve problems using **cubic inches**. Many everyday items are measured in **cubic units**. For example, the space in refrigerators and freezers is measured in cubic feet, and concrete is measured in cubic yards. You will find out what these terms mean and how they are used to measure volume.

You measured perimeter with linear units such as inches, feet, and centimeters. What units did you use to describe area? List some examples.

Volume is measured with a different type of unit. These units are called "cubic units," such as cubic inches, cubic centimeters, and cubic feet. Drawn below is a cubic centimeter.

Draw a cubic inch next to it. Label the dimensions: length, width, and height.

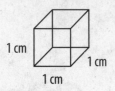

1 cm
1 cm
1 cm

Here is how to make a rectangle or square look three-dimensional:

- Draw the rectangle or square.

- Draw another one exactly like it, a little above and to the side of the first.

- Connect the top left corners, the top right corners, the bottom left corners, and the bottom right corners.

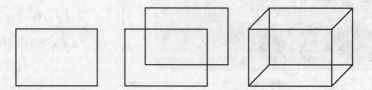

Activity 1: Inside the Wrapper

Using the **net**, construct a one-inch cube. Cut along the dark lines, fold on the light lines, and use tape to make the cube.

Using one of the wrappers (#1, #2, or #3), determine how many cubic inches fill the wrapper. How did you figure this out?

Activity 2: Cheese for a Party

You and your partner(s) will work on two different problems. First, you will figure out the number of cheese blocks needed for a party. Second, you will decide if a bulk-buying scheme makes sense.

1. Cheese, Anyone?

You are throwing a party for 20 people (including yourself). You plan to serve cheese and crackers as the main appetizer. You will cut your cheese into cubic inches, pieces small enough for guests to handle.

The cheese you are using comes in blocks that measure 3″ by 3″ by 4″. If you want to have enough cubes for each person at the party to have nine pieces, how many blocks of cheese will you need to buy?

- Show all work below. Use extra paper if needed.

- Label all numbers and pictures you use to solve the problem.

- Describe the steps you followed to solve the problem.

- How many blocks of cheese do you need to buy? _____

2. Volume Savings?

Your friend CJ likes to save money by buying in bulk. When he hears about your party, he suggests that you go to the Savings Warehouse and buy one big block of cheese to save money.

"How big a block would I have to buy?" you ask.

"Well, from the figures you have given me," he replies, "it looks like you will have to buy a 15″ by 15″ by 20″ block."

"That seems pretty big," you say.

"Oh, no, it's just what you need."

Do the math. Determine whether CJ is right or wrong, and prove it using pictures, words, and numbers. If you think he is wrong, what should the dimensions of the cheese block be? Keep track of your reasoning below (or on another sheet). Label all diagrams with *numbers* for the dimensions and *titles* such as "Original Cheese Block."

CJ is right because ...

or

CJ is wrong because ...

Practice: Pack That Ice Cream

Odie *loves* ice cream. In fact, with the warm weather coming, Odie has emptied two shelves of her freezer and plans to use the space to store as many packs of ice cream as she can.

- Each freezer shelf measures 12″ by 6″ by 24″.

- Each block of ice cream measures 6″ by 6″ by 4″.

Use diagrams, numbers, and words to answer the question: How many blocks of ice cream can Odie fit onto her two freezer shelves?

Show two solutions.

Practice: Comparing Volumes

Your teacher will provide you with copies of nets for two boxes. Cut, fold, and tape together Box A and Box B. Do not worry about the tops for the boxes just yet.

- Compare the volumes of the two boxes in cubic centimeters to determine whether one box has more room in it than the other.

- Show all your work below or use a separate piece of paper.

- Label all numbers and pictures you use to solve the problem.

- Describe all the steps you use to solve the problem.

- Indicate whether one box has more room than the other, and explain how you know.

Practice: Missing Dimension

Find the missing dimension for each of the following boxes.

1.

3 cm

3 cm

Volume = 27 cu. cm   Missing dimension: _____

2.

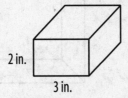

2 in.

3 in.

Volume = 24 cu. in.   Missing dimension: _____

3.

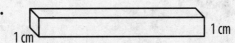

1 cm   1 cm

Volume = 9 cu. cm   Missing dimension: _____

4.

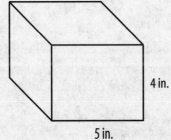

4 in.

5 in.

Volume = 100 cu. in.   Missing dimension: _____

Extension: A Special Box

The Acme Packaging Factory has been asked to design a special box (shown below). The entire box is 2" deep. What is the actual volume of the box?

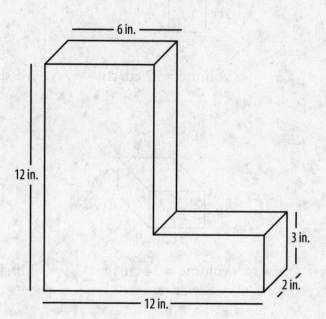

Test Practice

1. A cubic inch measures

   (1) 1′ x 1′ x 1′

   (2) 1′ x 1′

   (3) 1″ x 1″

   (4) 1″ x 1″ x 1″

   (5) 1″ + 1″ + 1″

2. The floor in a closet has an area of 40 sq. ft. If the closet has 320 cu. ft. of storage space, how tall is it?

   (1) 8 ft.

   (2) 10 ft.

   (3) 40 ft.

   (4) 80 ft.

   (5) 280 ft.

Question 3 refers to this net of an open box.

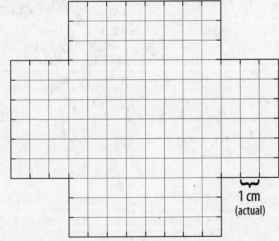

1 cm
(actual)

3. How many cu. cm could this box hold?

   (1) 28 cu. cm

   (2) 75 cu. cm

   (3) 90 cu. cm

   (4) 144 cu. cm

   (5) 160 cu. cm

4. Harry has a box that is 2″ wide, 4″ long, and 1″ high. How many cubic inches fit in Harry's box?

   (1) 6 cubic inches

   (2) 7 cubic inches

   (3) 8 cubic inches

   (4) 9 cubic inches

   (5) 10 cubic inches

5. Jan was stacking slabs of pine that measured 8′ by 5′ by 1′. Her forklift can carry seven of these slabs at once. Which expression below shows the total volume of slabs that Jan's forklift can carry?

   (1) 56′ x 35′ x 1′

   (2) 56′ x 35′ x 7′

   (3) 8′ x 5′ x 1′

   (4) 8′ x 35′ x 7′

   (5) 8′ x 5′ x 7′

6. 270 cu. ft. = _____ cu. yd.

On the Surface

How much cardboard
did it take to make
these boxes?

Boxes can have different shapes but the same volume. In this lesson, you will examine different boxes to find their volumes and their **surface areas.** You will then compare the amount of cardboard used to create different boxes. You will have an opportunity to explore the surface area of a box and compare it to the volume.

This lesson gives you a chance to practice using the volume formula.

Activity 1: Surface Area of a Box

1. How much cardboard or paper is used for the cheese wrapper?

    *   Examine the cheese wrapper.

    *   Record its dimensions: length, width, height, and volume.

    *   Find the number of square inches on the surface of the wrapper—its surface area.

    *   Record the surface area of the wrapper.

    Length: $l = $ _____   Width: $w = $ _____

    Height: $h = $ _____   Volume: $V = $ _____

    Surface Area: $S = $ _____

2. How much cardboard or paper is needed for this box?

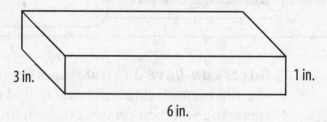

    *   Look at the sketch of the box.

    *   Record its dimensions: length, width, height, and volume.

    *   Record the surface area of the box.

    Length: $l = $ _____   Width: $w = $ _____

    Height: $h = $ _____   Volume: $V = $ _____

    Surface Area: $S = $ _____

Activity 2: Cardboard Needed

Directions

- Examine your group's box.

- Record its dimensions: length, width, height, and volume.

- Find the number of square centimeters on the surface of the box—its surface area.

- Record the surface area of your box.

- Post your findings on the chart at the front of the room.

Length: $l = $ _____

Width: $w = $ _____

Height: $h = $ _____

Volume: $V = $ _____

Surface Area: $S = $ _____

Look at the chart at the front of the room.

1. Write one sentence that describes what you noticed about the volumes of the boxes.

2. Write one sentence that describes what you noticed about the surface areas of the boxes.

3. Which box requires the most cardboard to make? How do you know?

Practice: Surface Area vs. Volume

Consider two gift boxes, Box A (12 cm x 5 cm x 3 cm) and Box B (6 cm x 10 cm x 3 cm).

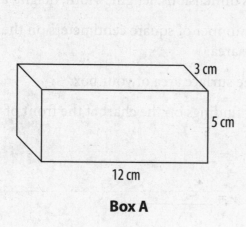

3 cm

5 cm

12 cm

Box A

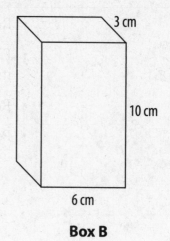

3 cm

10 cm

6 cm

Box B

1. How many cubic centimeters are in each of the gift boxes?

   Box A: _____

   Box B: _____

2. Which box is cheaper to make, Box A or Box B? How do you know?

Practice: Small Box Volumes

Boxes come in different sizes. The boxes for this problem are all small.
Their dimensions and volumes differ. Keep track of your box sizes in the
chart below. The five boxes, "open flat," are shown on pages 139–140.
The boxes are open. They do not have covers.

1. Think ahead and predict the following:

   a. Which boxes might be the largest in volume? _____

   b. Which boxes might be the smallest in volume? _____

   c. Which boxes might have the same volume? _____

2. Now measure and record your measurements below.

Box	Length cm (l)	Width cm (w)	Height cm (h)	Volume cubic cm (V)
A				
B				
C				
D				
E				

3. Which boxes had the same volume?

4. How accurate were your predictions?

5. Explain with drawings and words how boxes with different dimensions can have the same volume.

6. What size box would hold all these boxes? Show all work and explain your reasoning.

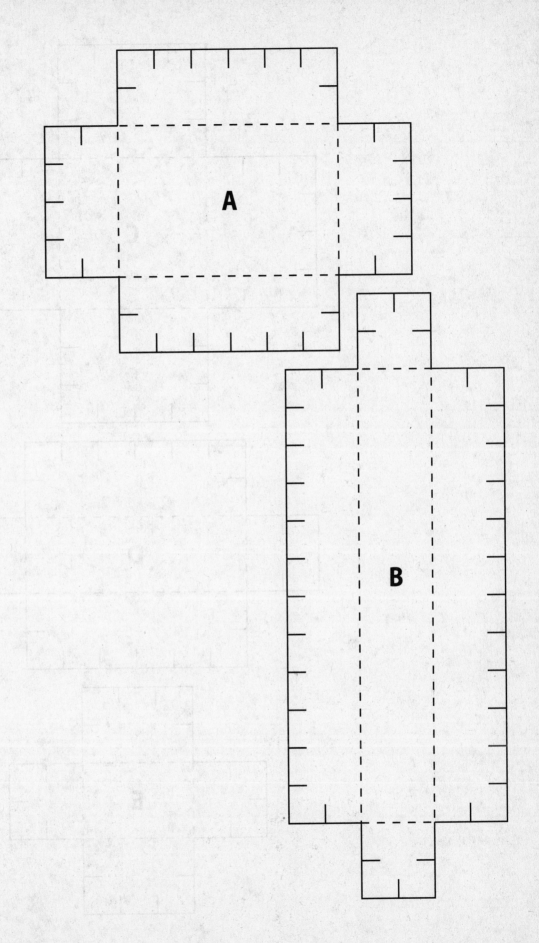

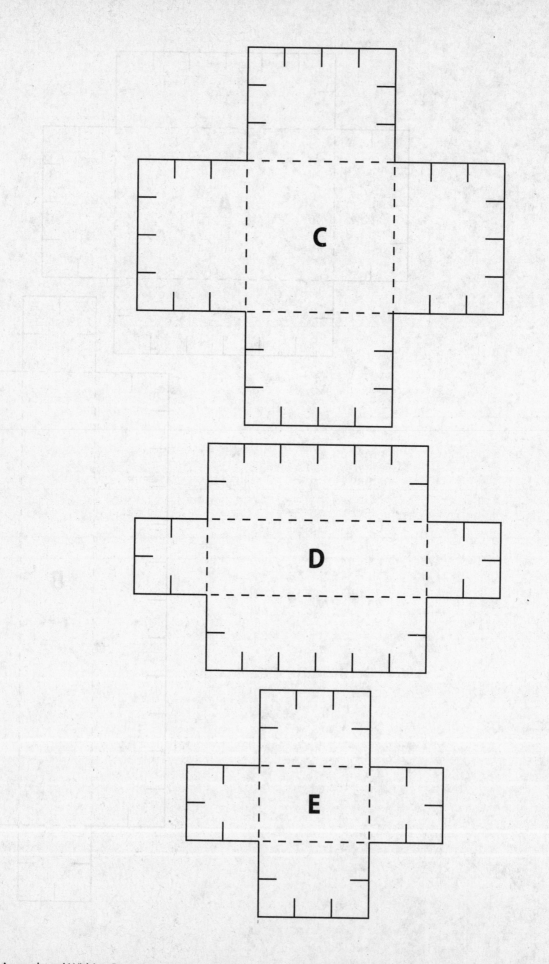

Practice: Vocabulary Review

Give a definition for or use in a sentence each of the words you know. Circle the ones you have seen but cannot define. If you have no idea, review your notes.

Parallel

Right angle

Perpendicular

Square

Perimeter

Area

Dimensions

Inch

Square inch

Volume

Isosceles triangle

Convert

Net

Yard

Centimeter

Scale

Hexagon

Cubic inch

Extension: Concrete Solutions

People who build and pour concrete foundations charge by the cubic yard. A cubic yard measures 3' by 3' by 3'.

1. How many cubic feet are in one cubic yard? Show your work below.

2. How many cubic feet are in three cubic yards? How do you know?

3. How many cubic yards are in 108 cubic feet? How do you know?

4. The Lane Company pours a foundation for the building below. The foundation is three feet deep. How many cubic yards do they use?

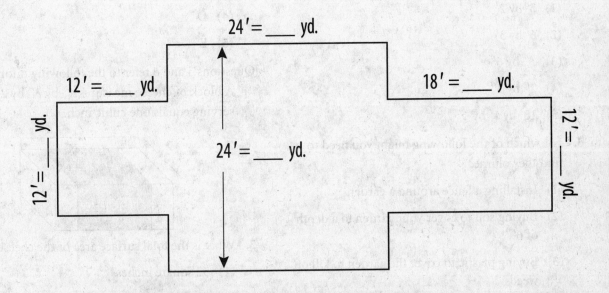

Remember that rectangles can be combined in a single shape.

Test Practice

1. Jana just bought a 24-cubic-foot refrigerator. Which expression below might represent the inside dimensions of her new appliance?

   (1) 2' x 2' x 12'

   (2) 3' x 3' x 4'

   (3) 3' x 3' x 6'

   (4) 2" x 3" x 4"

   (5) 2' x 2' x 6'

2. Charlene wants to know if the refrigerator she plans to buy will fit into her kitchen space. She knows the refrigerator is an 18-cubic-foot model and that it is 3 feet deep. What might the remaining dimensions measure?

   (1) 3' by 2'

   (2) 6' by 3'

   (3) 3" by 6"

   (4) 2" by 3"

   (5) 18' by 1' by 1'

3. For which of the following might you need to find the volume?

   (1) Installing a fence around a garden

   (2) Buying soil to cover your garden to a depth of 6"

   (3) Buying plastic to cover the garden to kill off weeds

   (4) All of the above

   (5) None of the above

4. All shapes below have the same volume. Which one would you expect to have the greatest surface area?

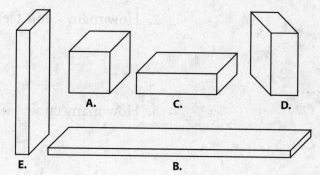

   (1) A

   (2) B

   (3) C

   (4) D

   (5) E

Questions 5 and 6 refer to the following information: A block of cheese measures 4" by 3" by 2". Each serving equals one cubic inch.

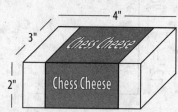

5. What is the total surface area of the cheese block?

   (1) 52 square inches

   (2) 52 cubic inches

   (3) 40 square inches

   (4) 40 cubic inches

   (5) 24 cubic inches

6. How many cubic inches are in the block of cheese?

Closing the Unit: Design a Box

> **Which box would be best?**

This session will give you another chance to work with volume. Every year one out of five adults in the United States moves. With all the packing and unpacking moving involves, it is important to have boxes that fit the items in your household. In this session you will think of an item and design the right-size box to fit four of your items.

Activity 1: Design a Box

1. What item are you packing?

2. How would you describe its dimensions?

3. What size box do you recommend to pack four of these items?

4. How much material (surface area) does the box require?

Activity 2: Review Session

1. Review what you learned in the unit.

   • Go back to the practice pages in past lessons.

   • Pick a page in each lesson.

   • Cover up what you wrote on the page.

   • Read the question only.

   • Answer out loud or on paper.

   • Reread your original answers to refresh your memory and check your work.

_____ Describe and name shapes	*Lesson 1*
_____ Describe and measure angles	*Lessons 2* and *3*
_____ Find area and perimeter of rectangles	*Lessons 4*, *5*, *6*, and *7*
_____ Measure and convert among inches, feet, yards, square inches, square feet, and square yards	*Lessons 8* and *9*
_____ Enlarge and draw to scale	*Lessons 4* and *10*
_____ Find the volume of rectangular solids using the formula *l* x *w* x *h*	*Lessons 11*, *12*, and *13*
_____ Compare surface area and volume	*Lesson 13*

2. Making a portfolio of your best work will help you review the concepts and skills in this unit (see page 159 for ideas).

Activity 3: Make a Mind Map

Make a Mind Map using words, numbers, pictures, or ideas that come to mind when you think of *Geometry* and *Measurement*.

GEOMETRY

MEASUREMENT

Activity 4: Final Assessment

Complete the tasks on the *Final Assessment*.

VOCABULARY

Lesson	Terms, Symbols, Concepts	Definitions and Examples
1	sides	
	corners	
	parallel lines	
	square	
	rectangle	
	triangle	
	hexagon	
	trapezoid	
2	acute angle	
	angle	
	right angle	
	obtuse angle	
	perpendicular lines	
3	straight angle	
	right triangle	
	isosceles triangle	
	equilateral triangle	
	scalene triangle	

VOCABULARY *(continued)*

LESSON	TERMS, SYMBOLS, CONCEPTS	DEFINITIONS AND EXAMPLES
4	length	
	similar shape	
	perimeter	
	width	
5	area	
	dimensions	
6	composite shape	
	square centimeter	
8	convert	
	inch	
	foot	
	yard	
9	square inch	
	square foot	
	square yard	
10	scale	

EMPower™

VOCABULARY *(continued)*

LESSON	TERMS, SYMBOLS, CONCEPTS	DEFINITIONS AND EXAMPLES
11	height	
	volume	
12	cubic inch	
	cubic unit	
	net	
13	surface area	

REFLECTIONS

OPENING THE UNIT:

_____
_____
_____
_____
_____
_____

LESSON 1: Sharing Secret Designs

What did you learn about shapes?

_____
_____
_____
_____
_____
_____
_____

LESSON 2: Get It Right

What do you want to remember about angles?

_____
_____
_____
_____
_____

What do you still wonder about?

LESSON 3: Get It Straight

What did you learn about straight angles?

What did you learn about triangles?

What do you still wonder about?

LESSON 4: Giant-Size

What do you want to remember about similar shapes?

_____

_____

_____

_____

_____

_____

LESSON 5: Line Up by Size

Use words and pictures to explain the difference between area and perimeter.

_____

_____

_____

_____

_____

_____

_____

LESSON 6: Combining Rectangles

What do you want to remember about square centimeters?

_____

_____

_____

_____

_____

_____

_____

LESSON 7: Disappearing Grid Lines

What do you want to remember about finding missing information about rectangles?

_____

_____

_____

_____

_____

What do you still wonder about?

_____

_____

_____

_____

_____

LESSON 8: Conversion Experiences

What do you want to remember about converting among units?

_____

_____

_____

_____

_____

What do you still wonder about?

_____

_____

_____

_____

_____

LESSON 9: Squarely in English

What is your understanding of area and how it is measured? Use drawings and words to explain.

_____

_____

_____

_____

_____

_____

What do you want to remember about converting square feet to square inches?

_____

_____

_____

_____

_____

_____

What do you want to remember about converting square yards to square feet?

_____

_____

_____

_____

_____

What do you want to remember about converting square yards to square inches?

_____

_____

_____

_____

_____

_____

LESSON 10: Scale Down

What do you want to remember about drawing to scale?

_____

_____

_____

_____

_____

LESSON 11: Filling the Room

What did you have to keep in mind when filling the room with boxes?

_____

_____

_____

_____

_____

_____

LESSON 12: Cheese Cubes, Anyone?

What do you know about volume?

_____

_____

_____

_____

_____

What is important to remember about cubic units?

_____

_____

_____

_____

LESSON 13: On the Surface

What do you want to remember about surface area and volume?

_____

_____

_____

_____

_____

_____

CLOSING THE UNIT

Your Best Work

Review work you have done in class and on your own. Pick out two assignments you think are your best work or show where you learned the most.

Make a cover sheet that includes

- Your name
- Date
- Names of the assignments

For each assignment that you pick

- Write a sentence or two describing the piece of work.
- Write a sentence or two explaining what skills were required to complete the work.
- Write a sentence or two explaining why you picked this piece of work.

> ## Take some time to look through your work.
>
> What did you do?
>
> What did you learn?
>
> Look back at your *Reflections* and *Vocabulary* to get more ideas.

SOURCES AND RESOURCES

Web Sites

For moveable pattern block shapes online, see
http://arcytech.org/java/patterns/patterns_j.shtml

For pattern blocks, see
http://www.best.com/~ejad/java/patterns

For answers to geometry-related questions, see
http://mathforum.org/dr.math/

For a geometry reference resource for children and adults, see
http://w3.byuh.edu/library/curriculum/Geometry/Geometry.htm

To check estimates for *A Fresh Look* using Home Depot's online
calculator, see http://www.homedepot.com

For information on fractal geometry, see
http://www.strengthinperspective.com/MPART/MPGAL5/mpgal5.html

Geometry and Art

http://www.imart.org/education/IMA_education/volume.asp?levelid=2

http://www.isibrno.cz/~gott/mandalas.htm

http://www.ed.uri.edu/SMART96/ELEMATH/GeometryArt/geometry.html

http://www.myschoolonline.com/content_gallery/0,3138,52947-130799-
56-6394,00.html

http://www.cs.berkeley.edu/~sequin/SCULPTS/

http://mathforum.org/~sarah/shapiro/

http://www.cs.berkeley.edu/~sequin/ART/

http://www.lotuslazuli.com/contents.htm

http://www.mathcs.carleton.edu/penrose/index.html